Cities and Nature

Series Editors
Peter Newman, Sustainability Policy Institute, Curtin University, Perth, WA, Australia
Cheryl Desha, School of Engineering and Built Environment, Griffith University, Nathan, QLD, Australia
Alessandro Sanches-Pereira, Instituto 17, São Paulo, São Paulo, Brazil

Cities and Nature fosters high-quality multi-disciplinary research addressing the interface between cities and the natural environment. It provides a valuable source of relevant knowledge for researchers, planners and policy-makers. The series welcomes empirically based, cutting-edge and theoretical research in urban geography, urban planning, environmental planning, urban ecology, regional science and economics. It publishes peer-reviewed edited and authored volumes on topics dealing with the urban and the environment nexus, including: spatial dynamics of urban built areas, urban and peri-urban agriculture, urban greening and green infrastructure, environmental planning, urban forests, urban ecology, regional dynamics and landscape fragmentation.

Zaheer Allam • Peter Newman

Revising Smart Cities with Regenerative Design

Springer

Zaheer Allam
Curtin University Sustainability Policy Institute (CUSP)
Curtin University
Perth, WA, Australia

Peter Newman
Curtin University Sustainability Policy Institute (CUSP)
Curtin University
Perth, WA, Australia

ISSN 2520-8306 ISSN 2520-8314 (electronic)
Cities and Nature
ISBN 978-3-031-28027-6 ISBN 978-3-031-28028-3 (eBook)
https://doi.org/10.1007/978-3-031-28028-3

This Springer imprint is published by the registered company Springer Nature Switzerland AG
The registered company address is: Gewerbestrasse 11, 6330 Cham, Switzerland

Preface: Into the Maelstrom

We are writing a book on Smart Cities at a remarkable time in history. The Smart City idea has taken centre-stage in much of the urban agenda for the past 20 years. It expressed what people wanted from a modern city. It was a competition to see who the smartest city could be and, as we will assert, to a large degree it sapped the life and money out of the other deeper agendas of sustainability, climate resilience and regenerative design. But that has all changed, we are now in the maelstrom of creating a completely new kind of city.

An illustration of this can be provided before moving into the book in detail to more specifically engage with how the Smart City concept needs to be redefined. In Toronto a waterfront project called Quayside was put out to tender and the winner was a Smart City project called Sidewalk Toronto. The project set out to demonstrate the ultimate in Smart City technology with a primary function of digital data collection to assist urban planning. However, the project stumbled because it was never clear how the data would be used for other than private benefit for the major corporation (Google) behind the development. As described by Bianca Wylie, the fundamental problem of this project is the Smart City model itself. She argues that the model is formulated because 'corporations are seeking to exert influence on urban spaces and democratic governance (Wylie, 2018)'.

Strong opposition grew as the public had to assess the idea that Toronto was going to become a hub for an optimsed urban experience featuring robo-taxis, heated sidewalks, autonomous garbage collection and an extensive digital layer to monitor everything from street crossings to park bench usage. The ultimate Smart City was being proposed based on large increases in digital data capacity.

In the end, the project was abandoned, and an alternative proposal was generated that pushed other values than just being smart. The alternative featured ideas that included a much greater element of nature-positive wildness in its biophilic character. In particular, it featured net zero outcomes. In an article called 'Kill the Smart City' from *MIT Technology Review*, Karrie Jacobs (2022) says 'People are attracted to the messiness of a wild and diverse city…not to be quantified and controlled'. She continued 'Smart city technology should improve transit and reduce carbon emissions…so Quayside shows trees and greenery coming from every balcony with nary an autonomous vehicle or drone in sight'.

Figure 1 shows why it is a different kind of city that is being dramatised than the Smart City option originally considered. It is a battle of ideas that has clearly stated to the Smart City concept: you need to be redefined.

What is not explained in the *MIT Technology Review* article is that in order to achieve these goals there would need to be a very significant dependence on Smart City technology. The difference is that these technologies would be subservient to the broader environmental, social and climate values that must now be driving the development process.

This is the core idea of our book – we need the Smart City concept to be quickly redefined so it can help us solve the issues of sustainability, climate resilience and other major UN agendas suggested by the Sustainable Development Goals (SDGs). And to do this we need to use regenerative design.

We need to do this quickly because we have entered the maelstrom of post-COVID-19 economic change and it needs a lot of help.

The website Climate 100+ contains all the finance companies now committed to only funding Net Zero projects and other SDGs. There is over $170 trillion that is committed to this cause in 2022 which although this was part of the Paris Agreement from 2015 it has only begun to accelerate since the post-COVID-19 rebuilding of the economy. As will be explained in this book, the Net Zero project in cities can

Fig. 1 The concept for the Toronto Waterfront which killed off the Smart City concept previously awarded

Preface: Into the Maelstrom

We are writing a book on Smart Cities at a remarkable time in history. The Smart City idea has taken centre-stage in much of the urban agenda for the past 20 years. It expressed what people wanted from a modern city. It was a competition to see who the smartest city could be and, as we will assert, to a large degree it sapped the life and money out of the other deeper agendas of sustainability, climate resilience and regenerative design. But that has all changed, we are now in the maelstrom of creating a completely new kind of city.

An illustration of this can be provided before moving into the book in detail to more specifically engage with how the Smart City concept needs to be redefined. In Toronto a waterfront project called Quayside was put out to tender and the winner was a Smart City project called Sidewalk Toronto. The project set out to demonstrate the ultimate in Smart City technology with a primary function of digital data collection to assist urban planning. However, the project stumbled because it was never clear how the data would be used for other than private benefit for the major corporation (Google) behind the development. As described by Bianca Wylie, the fundamental problem of this project is the Smart City model itself. She argues that the model is formulated because 'corporations are seeking to exert influence on urban spaces and democratic governance (Wylie, 2018)'.

Strong opposition grew as the public had to assess the idea that Toronto was going to become a hub for an optimsed urban experience featuring robo-taxis, heated sidewalks, autonomous garbage collection and an extensive digital layer to monitor everything from street crossings to park bench usage. The ultimate Smart City was being proposed based on large increases in digital data capacity.

In the end, the project was abandoned, and an alternative proposal was generated that pushed other values than just being smart. The alternative featured ideas that included a much greater element of nature-positive wildness in its biophilic character. In particular, it featured net zero outcomes. In an article called 'Kill the Smart City' from *MIT Technology Review*, Karrie Jacobs (2022) says 'People are attracted to the messiness of a wild and diverse city…not to be quantified and controlled'. She continued 'Smart city technology should improve transit and reduce carbon emissions…so Quayside shows trees and greenery coming from every balcony with nary an autonomous vehicle or drone in sight'.

Figure 1 shows why it is a different kind of city that is being dramatised than the Smart City option originally considered. It is a battle of ideas that has clearly stated to the Smart City concept: you need to be redefined.

What is not explained in the *MIT Technology Review* article is that in order to achieve these goals there would need to be a very significant dependence on Smart City technology. The difference is that these technologies would be subservient to the broader environmental, social and climate values that must now be driving the development process.

This is the core idea of our book – we need the Smart City concept to be quickly redefined so it can help us solve the issues of sustainability, climate resilience and other major UN agendas suggested by the Sustainable Development Goals (SDGs). And to do this we need to use regenerative design.

We need to do this quickly because we have entered the maelstrom of post-COVID-19 economic change and it needs a lot of help.

The website Climate 100+ contains all the finance companies now committed to only funding Net Zero projects and other SDGs. There is over $170 trillion that is committed to this cause in 2022 which although this was part of the Paris Agreement from 2015 it has only begun to accelerate since the post-COVID-19 rebuilding of the economy. As will be explained in this book, the Net Zero project in cities can

Fig. 1 The concept for the Toronto Waterfront which killed off the Smart City concept previously awarded

only be achieved if the parallel concept of Smart Cities is also being applied. This is detailed in the book, but the need for Net Zero is rapidly becoming a part of every facet of city life.

So why is it a maelstrom? The Climate 100+ website does not suggest that the world of finance knows all the answers, but it is motivated to see the changes in the next economy which have been talked about for too long, being started now. Thus, the website's front page is a waterfall as in Fig. 2.

The image is suggesting that climate action may well have been growing faster over many years, like a slowly accelerating river, but now we are going over the edge into the maelstrom. Rapid change is happening, and no one is quite sure how to proceed. Money is flowing in unprecedented amounts into how to make our cities and regions decarbonised and there are no obvious manuals of how to proceed. This book is suggesting that we are indeed in a maelstrom of change but if we are to reach the quiet waters of the downstream economy, we will need to link Smart City technology into every part of the new green, Net Zero economy.

Although the Smart City concept has been a dominant factor in cities over the past two decades, it has never been seen by UN agencies as having any more legitimacy than a branding exercise and competition among big digital economy corporations. This has now begun to ease as its value in many UN agendas is growing and the opportunity is even more obviously mainstreaming due to Net Zero mandates and the other 17 Sustainable Development Goals which are always associated with any plans for the future. They are all beginning to see a new role for Smart City technology and systems. But Smart City providers – companies, city managers, and government funding agencies – need to see ways to design and procure the smart

Fig. 2 The maelstrom of a waterfall symbolising what global investors suggest is the transition period we are entering

technologies as an integrated part of the big agendas being set out by the UN and now being backed by Climate 100+ financiers.

The journey in this book is to show some of the history in how Smart City has grown as a concept and how it has been understood and misunderstood. How it became an agenda that substantially blocked the other big agendas of sustainability and climate resilience.

We suggest Smart City needs to be redefined and integrated into regenerative design to make its true message and value enabled, to guide us through the maelstrom. We will set out what this can mean as we apply it to the UN's big topics of sustainability and climate change and also economic development that enables inclusion, emphasising the critical role of Smart City technology in integrating solutions into the delivery process.

We finalise the book by explaining how the emerging concept of regenerative design can help us more deeply understand how to get through the maelstrom. We suggest that the regenerative approach will enable us to position Smart Cities as a major part of the development agenda, in low- and high-income cities.

Perth, WA, Australia Zaheer Allam

Peter Newman

Contents

Chapter 1
Unveiling the Smart City: How Smart Is It?

Abstract The emerging concept of Smart Cities brings about promises of increased efficiency and performance of urban areas through the use of specialised digital technology. With an aim to promote innovation, this is prompting a wide adoption in high- and low-income economies as most countries embark on strategies to use the concept to boost foreign investment and financial confidence and to showcase national innovation. However, as the demand for the technology inherent in Smart Cities booms, questions arise as to whether the concept is promoted primarily by ICT corporations driven by profit-making and merely equates supply with demand without any other fundamental values for creating a better future. We suggest that Smart City technology needs to be driven by these deeper values and be integrated into delivery of solutions to multiple local and global needs. This chapter explores this conundrum and showcases the need for tailored solutions rather than "off-the-shelf" technology, as is mostly offered by ICT corporations, and outlines how deeper values as set out through UN processes about sustainability and climate resilience are now essential components of how Smart City is imagined. It also introduces the concept of regenerative design that will be needed to guide how Smart City technology is procured and delivered in the future as an integrated approach to the future city.

1.1 Introduction

The Smart City concept has gained traction and support throughout the world and is now considered as a global phenomenon. The concept only started around 1997, as advanced by Graham and Aurigi (1997). Before this, scientists and engineers from different parts of the world are documented to have only been simulating virtual cities, mirroring most of the concepts known to be associated with today's Smart City concepts. Anthopoulos (2017) supports that this drive to run simulations to further explore alternative solutions was driven by factors like lack of green spaces, violence and insecurity and perceived reduced civil interactions, among others. Such simulation of virtual cities was enabled by the World Wide Web (WWW) and the spread of Internet such that people could make use of devices to interact

Z. Allam, P. Newman, *Revising Smart Cities with Regenerative Design*, Cities and Nature, https://doi.org/10.1007/978-3-031-28028-3_1

virtually. Those advances, through the virtual city concept, helped to steer transformation toward a digitally oriented city, as advanced by van den Besselaar and Beckers (1998), and unlike the virtual city, it was seen to promote the need for space to enable social and human interaction, and since the idea was actualized via the Internet, it provided the opportunity for an increased interaction beyond the limitations of physical locality. The virtual and digital city concepts were furthered with the integration of urban management centres in the form of digital platforms for local administration and for citizens to interact, especially for information collection and sharing. Anthopoulos (2017) suggests that promoters of these two concepts that are akin to today's Smart Cities perceived urban spaces as interlinked 'islands of communities' that provided those living in them the opportunities to benefit from services and information accessed via the Internet. Ishida (2017) suggests that the concept of digital city was launched in Kyoto in 1998 and provided the opportunity for human interactions to be captured via cameras and simulations that were projected in the form of animations resulting in 2D and 3D virtual spaces.

Other terminologies have been used over the years that can be viewed to have helped create the Smart City concept. Such terminologies like the information city, ubiquitous city and intelligent city have been, in one way or the other, used to describe a city that incorporates the power of information communication technology (ICT) in its core activity. Anttiroiko et al. (2014) believe that the practice of Smart Cities initially started as the virtual city, which has been developed in a more sophisticated manner.

The range of technologies that today are associated with a more sophisticated and complex idea of Smart Cities have emerged from ICT technologies and consist of Internet of Things (IoT), big data, crowdsourcing, machine learning, artificial intelligence, digital twins, mobile connectivity (e.g. 4G and 5G) and blockchain technologies, among others (Allam 2018a, c; Allam and Dhunny 2019; Huiling and Goh 2017). Sepasgozar et al. (2018) claim that these technologies have become the cornerstone of Smart Cities since they offer platforms for installation of 'smart' components and infrastructures and offer systems that allow for real-time collection of big data and analysis of the same. The different varieties and increasing quantity of smart devices and sensors installed in cities, coupled with the advancement in social networks provided by ICT, allow city managers and other stakeholders to gather, analyse and react to emergent data. Thus, real-time action can be analysed to enable better efficiency, speed, scalability and flexibility in almost any kind of urban activity.

It is not surprising then to see that these 'smart' technologies could or even should play a significant role in transforming the design, planning and management of urban life, especially in addressing issues related to healthcare, traffic, communication, environmental sustainability and economic growth. These issues have occupied us for most of the past decade as we tried to define whether Smart City was doing all that it promised or whether it needed to be redefined (Allam 2018b; Allam and Jones 2018a, b; Allam and Newman 2018a; Khan et al. 2017).

Despite the notable benefits and promises of the Smart City concept, there are some reservations demonstrated by both individuals and organisations, like the

United Nations (UN) (Allam 2018a; Allam and Newman 2018b). It is important to see that the response by the UN to this concept has been very guarded. The global debate about future cities has many dimensions and contributors, and there was much written about the importance of the United Nations Sustainable Development Goals for 2015–2030, which now include an urban goal SDG11: 'inclusive, safe, resilient and sustainable cities'. This goal (SDG11) has 10 targets and 14 indicators, but it is of interest that in all these words about what cities need to do, none are saying that we should have Smart Cities, despite the increasing use of the term. The reason apparently is that Smart Cities are seen to be essentially a branding war between different multinational corporations in the ICT space. In a 2016 report by the Economic and Social Council (ECOSOC) of the UN (United Nations 2016), these reservations are pronounced. The report acknowledges that the Smart City concept has the potential to address issues related to urbanization trends and to help in achieving the Sustainable Development Goals (SDGs), but it raises some concerns. The challenges are seen to be tied to the implementation of the concept and include (1) localisation of smart infrastructures, (2) lack of skilled labour, (3) financing challenges, (4) application of a suitable governance model and (5) the challenge of inclusion. The report is particularly critical of the lack of inclusive models in the delivery of the Smart City.

Most critics of Smart City strategies that are being proposed are seen as being borrowed from other geographical locations and areas where the programs have apparently been successful; hence, they fail to address the local development challenges. By failing to integrate local solutions to Smart City programs, issues like security and privacy concerns become prominent, and social inequality is deemed to increase. For instance, despite the increased adoption of Smart City concepts in different geographies, cases of homelessness have been on the rise (Bezgrebelna et al. 2021). Such are attributed to factors like gentrification, prompted by impacts of application of 'smart' technologies in cities, making cost of living untenable for most of the residents (Wilhelmsson et al. 2021).

Though the concerns by the UN and others like Yigitcanlar et al. (2018), Mosenia and Jha (2017) and Alomair and Poovendran (2014) are valid and demand attention, the concept of Smart City is still very popular, and we too believe it has considerable potential (Allam and Newman 2018b; Braun et al. 2018). However, there is a particular need to redefine Smart City to enable it to reach its full potential.

The solution, as set out in this book, is for cities to take the goal, targets and indicators and see how they fit the kind of technological opportunities that are emerging as the Smart City. Perhaps in this way, the Smart City can be revised into being more than an economic tool in the tradition of modernist systems of change and rather stand as an accelerator for 'inclusive, safe, resilient and sustainable cities'.

1.2 The Adoption of Smart Cities

Even though Smart Cities are well used in popular and academic literature, a proper definition is still debatable (Albino et al. 2015; Chourabi et al. 2012). Table 1.1 sets out six definitions based mostly on the reviews done by Chourabi et al. (2012) and Cocchia (2014). Often the definitions are just what a ‘good’ city should be while others emphasise technology.

What is universally agreed is the potential use of the concept to solve numerous urban challenges where data analysis could help. However, numerous stakeholders have capitalised on the concept as a branding tool to encourage an increased adoption by numerous cities, especially for surveillance and security purposes. The perceived popularity of ‘Smart Cities’ in contrast to its less favourable counterparts like ‘sustainable cities’ and ‘resilient cities’ is probably because the proponents of smart technologies are not seriously addressing the issues of resilience and sustainability. This book seeks to address that issue.

The term sustainable cities emerged with the need for cities to address sustainable development (Satterthwaite 1997), while ‘resilient cities’ emerged with the question of planners and designers of how to quickly and efficiently recover from urban perturbations, often linked with climate change (Vale 2014). A comparative analysis (Fig. 1.1) of the terms ‘Smart Cities’, ‘sustainable cities’ and ‘resilient cities’ (Google 2018a, b) shows that the term sustainable cities was more popular until late 2010. Following this, the term Smart Cities emerged as most popular. In August 2015, the terms ‘resilient cities’ was factored at 3% and ‘sustainable cities’ at 5% in comparison with the popularity of ‘Smart Cities’, which was at its peak. This trend underlines questions as to how Smart Cities gained popularity as opposed to its counterparts.

Cocchia (2014) suggests that the keywords smart, intelligent, knowledgeable, sustainable, digital and ubiquitous have often been associated with cities that have embraced the use of ICT to address various urban issues. Those are in turn made to appear and have evoked a sense of ‘modernity’ which most cities, particularly low-income and emerging economy cities, aim to achieve. Beside the catchy keywords, the associated technologies that render improved urban fabrics provide an aura of modernity through order and synchronicity. Issues like traffic, housing, water and energy provision, security and the environment are thus possible to be addressed in this kind of branding. Those can even be made to positively impact on branding through cultural and artistic dimensions of cities and improve on its infrastructure, all with an aim of improving a societal sense of belonging (Allam 2018b). To et al. (2018) show this very clearly by asserting that smart infrastructures as included in high rise buildings, transportation routes, recreation centres and street lighting projects are not only proposed to make cities ‘smarter’ but are designed in such a way that they embrace ‘modernity’, which in turn helps to improve urban economic status. For instance, it has been noted that cities such as Singapore, Tokyo, Barcelona, Amsterdam and Copenhagen, among many others, have improved on their attraction by adopting the Smart City concepts (Gascó-Hernandez 2018; Joss et al. 2019;

Table 1.1 Proposed definitions of Smart City adapted from Chourabi et al. (2012) and Cocchia (2014)

Author/s	Definition
Giffinger et al. (2007)	A city well performing in a forward-looking way in economy, people, governance, mobility, environment, and living, built on the smart combination of endowments and activities of self-decisive, independent and aware citizens
Hollands (2008)	A city that monitors and integrates conditions of all of its critical infrastructures, including roads, bridges, tunnels, rails, subways, airports, seaports, communications, water, power, even major buildings, can better optimize its resources, plan its preventive maintenance activities, and monitor security aspects while maximizing services to its citizens
Harrison et al. (2010)	A city 'connecting the physical infrastructure, the IT infrastructure, the social infrastructure, and the business infrastructure to leverage the collective intelligence of the city'
Natural Resources Defense Council (2018)	A city striving to make itself 'smarter' (more efficient, sustainable, equitable, and livable)
Toppeta (2010)	A city 'combining ICT and Web 2.0 technology with other organizational, design and planning efforts to dematerialize and speed up bureaucratic processes and help to identify new, innovative solutions to city management complexity, in order to improve sustainability and liveability'
Washburn et al. (2009)	'The use of smart computing technologies to make the critical infrastructure components and services of a city—Which include city administration, education, healthcare, public safety, real estate, transportation, and utilities—More intelligent, interconnected, and efficient'
Setis-Eu (Cited in Cocchia 2014)	'Smart City is a city in which it can combine technologies as diverse as water recycling, advanced energy grids and mobile communications in order to reduce environmental impact and to offer its citizens better lives'
Dameri (2012)	'A Smart City is a well-defined geographical area, in which high technologies such as ICT, logistic, energy production, and so on, cooperate to create benefits for citizens in terms of well-being, inclusion and participation, environmental quality, intelligent development; it is governed by a well-defined pool of subjects, able to state the rules and policy for the city government and development'
Northstream (2010)	'Concept of a Smart City where citizens, objects, utilities, etc., connect in a seamless manner using ubiquitous technologies, so as to significantly enhance the living experience in twenty-first century urban environments'
Hall et al. (2000)	'A city that monitors and integrates conditions of all of its critical infrastructures, including roads, bridges, tunnels, rails, subways, airports, seaports, communications, water, power, even major buildings, can better optimize its resources, plan its preventive maintenance activities, and monitor security aspects while maximizing services to its citizens'
Su et al. (2011)	'Smart City is the product of Digital City combined with the internet of things'
IBM (2010)	'Smart City is defined by IBM as the use of information and communication technology to sense, analyze and integrate the key information of core systems in running cities'
California Institute (2001 cited in Cocchia (2014 #287))	'A smart community is a community that has made a conscious effort to use information technology to transform life and work within its region in significant and fundamental rather than incremental ways'

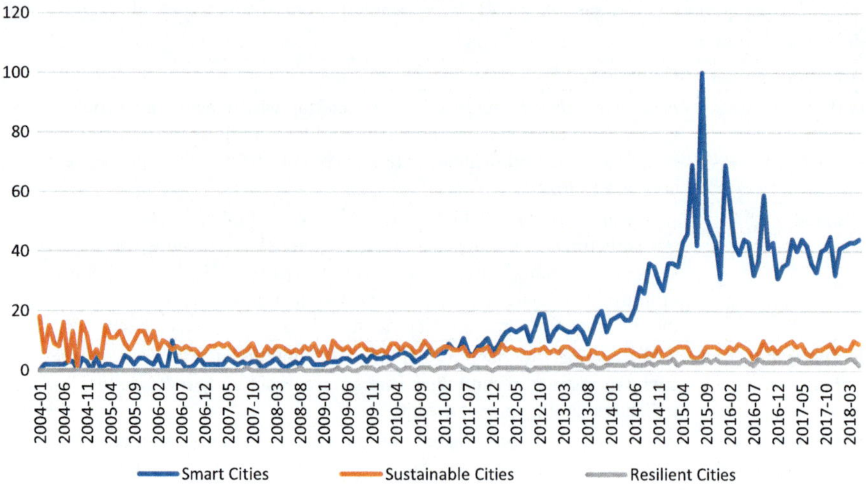

Fig. 1.1 Number of searches for three types of cities worldwide. (Google 2018a, b)

Rohaidi 2018; McKinsey & Company 2018a, b). Interestingly, a majority of these cities have managed to include sustainable development agendas which also helps in improving economic growth, as advanced by Trindada et al. (2017). But they are an afterthought, not the mainstream purpose which is to assert that digital data capacity can solve most problems.

The concept of Smart Cities has been gaining momentum around the world as set out in Figs. 1.2 and 1.3 though, as Fig. 1.2 suggests that it may have peaked in 2015.

The popularity of Smart Cities projects and programs has been mushrooming across the globe, such as in India, China, UAE, South Korea and even in small island developing states like Mauritius (Datta 2015; Glasmeier and Nebiolo 2016; Nam and Pardo 2011; Allam 2017; Kitchin 2014). Data from 2004 to 2018 was sourced from Google Trends (Trends 2018), and the y-axis on both Figs. 1.1 and 1.2 highlights the popularity (ranging from 0 to 100). A study of the term 'Smart Cities' surprisingly highlights that Smart Cities are most popular in Mauritius (Fig. 1.3), and a case study on these Smart Cities is presented below.

Kolotouchkina and Seisdedos (2017) explain that cities such as Songdo in South Korea and Masdar in Abu Dhabi, among others, have managed to tap into the perceived status of Smart Cities to rebrand their cities as attractive destinations based on place-branding strategies. However, over time, those have been unable to fulfil their original branding as eco-cities or Smart Cities. Smart city branding has been sought by many cities from low-income and emerging economies. Mancebo (2020) further suggests that the Smart City concept has become a successful branding tool that has allowed cities to maintain a competitive edge while highlighting their potential technological throughputs. Sofeska (2017) contends that this branding is advanced by the adoption of seemingly high-tech strategies aimed at the improvement of resilience and liveability status of cities, for the aim to rank higher in liveability

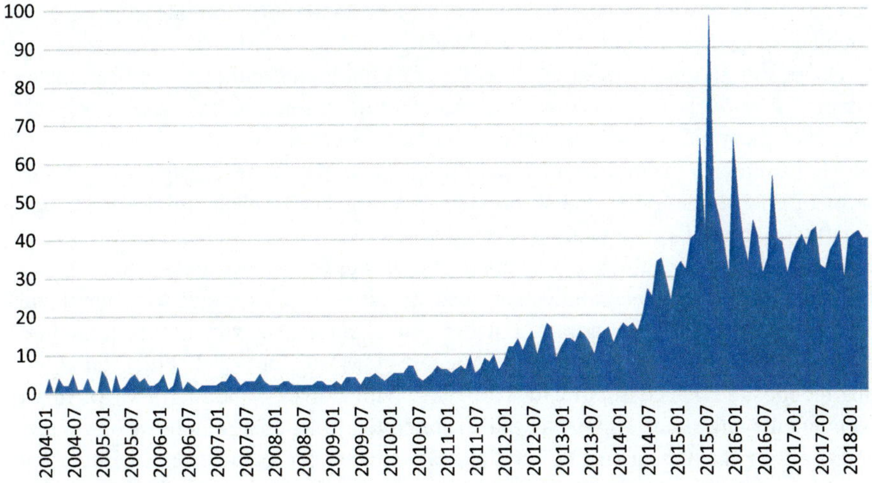

Fig. 1.2 Relative number of hits for Smart Cities searches in Google between 2004 and 2018. (Google 2018a, b)

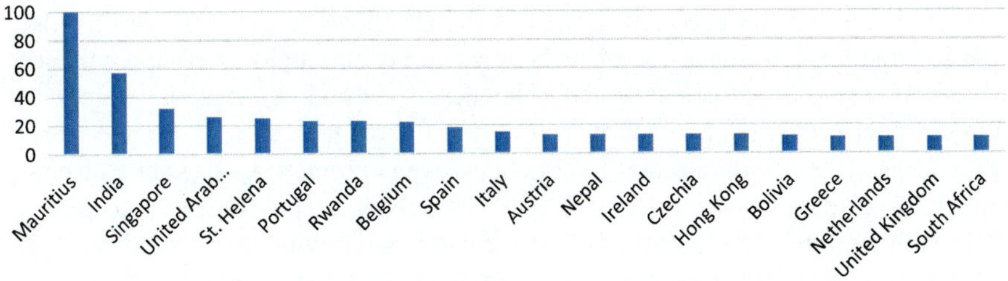

Fig. 1.3 Countries with most searches for Smart Cities between 2004 and 2018. (Source: Google Trends Explore)

status indexes. But there is little obvious direction as to how digital data capacity can actually do that in any of their strategies.

At the centre of this branding is the use of big data, automation and other technological advancements that are now synonymous with Smart Cities (Allam and Dhunny 2019). The data generated by the numerous and diverse number of installed 'smart' components across a range of networks are generally hoped to interact between each other. Indeed, technologies such as IoT (Bruneo et al. 2019; Bibri 2018), machine learning, cloud computing and even blockchain technologies (Barkham et al. 2018; Souza et al. 2016) can allow Smart Cities to automate different aspects of urban life in a coherent way, but these technological options must be given policy and strategic direction. The associated employment opportunities that have come from each of those smart technologies is used to further support its branding strategies but the ability to show each of these technologies can enable outcomes addressing the big UN agendas of sustainability, climate change economic development and inclusion, are rarely provided.

Huertas et al. (2021) dwell on the role of technologies and their applications to Smart City models in reinventing the character of different cities. They acknowledge that most cities, faced with the devastating impacts of climate change, have adopted strategies to optimize the use of resources and to better engage with dimensions of environmental sustainability, but they suggest no clear link to how these Smart City models can actually achieve this.

By using such technologies like big data analytics to improve areas that can positively impact on local economies and security, cities have managed to rebrand and reposition themselves to attract a wider global audience, which translates into increased visitors. This is captured in a report by Mastercard (2017) that found out that the most advanced Smart Cities are those with a diverse and multitude of smart components and smart systems in place, aimed to attract a large number of visitors and in aiding the management of urban areas. This is seen as increasingly important for urban policy makers as visitors contribute to a sizeable revenue from the consuming of different services and products on offer. The technologies work at enabling specific local areas to be branded as 'modern' and 'safe'. But is this enough?

The place-branding strategy that a majority of Smart Cities have embarked on to position themselves for economic development has allowed many cities, even those with low incomes, to make use of technology to create employment and enhance urban management, performance and efficiency. They hint that this may help with all their other problems, but in reality, they don't know how to do it. So, the world has witnessed the proliferation of Smart Cities, even in remote areas where the concept looks disparate and unnecessary. An annual report by Berrone and Ricart (2018) released in 2018 captures 165 Smart Cities worldwide, but a good score of them, as admitted in the report, has not been included in their analysis.

McKinsey and Company (2018a, b) suggested that the number of Smart Cities is set to continue to rise as technological advancements take shape. This book will show that unless the Smart City technology is made part of the bigger UN agendas, there will not be much further expansion. We believe that Smart City as a branding exercise for digital data capacity needs to diminish and the big agendas need to drive the future of cities. But we will also show that Smart City will be an essential part of that bigger agenda.

1.3 Smart Cities as a Profitable Venture

The discourse about the Smart City concept being used as a branding tool aimed at promoting economic growth through the technologies associated with it suggests that this demand is an incredibly profitable venture. McKinsey and Company (2018a, b) do show that most Smart City programs are powered by components, systems, organisations, businesses and individuals who aim to benefit financially from these undertakings. This is affirmed by Richter et al. (2015) who argue that even though small local entrepreneurs have failed to capitalize on the concept of Smart Cities, large corporations are investing massively in both development and branding of Smart City. The same argument is advanced by Taylor (2014), from CISCO, who shares how service providers can tap into the concept of Smart Cities to increase economic throughput from the provision of a wide and diverse array of 'smart' products and services.

This profit-seeking drive however can be made to the detriment of the end users. They can be delivered as unfair economic models that marginalize the big issues from the UN about sustainability and climate change. An example is from the transportation sector where companies such as Uber and Google capitalize on the improvement of public infrastructure (aided by public taxpayers) to improve their private profit targets. Large corporations such as IBM and Cisco have been in the forefront of such technologies and have made substantial profit and built their reputation from the sale of hardware, software and digital services, platforms and solutions. Their positioning have earned them the opportunities to services cities such as Rio, Barcelona, Kansas and others in countries that have national Smart City agendas like China and India (McKinsey & Company 2018a, b). However, at no stage have these projects linked in with strategies to increase electrification of transport or to provide solar energy as the fuel.

Other large companies like Google, ABB, General Electric, Ingersoll Rand, Siemens AG, Hitachi Ltd., Huawei Technologies Co. Ltd., Koninklijke Philips and Microsoft (Mordor Intelligence 2018) are also among the tech innovators and providers of Smart City technologies particularly on artificial intelligence (AI). None of these stand out as leaders in sustainability and climate mitigation, but they could have used this technology to help drive those options.

The Smart City approach to solving urbanisation issues is not a stand-alone concept and is backed and supported by corporates with substantial financial resources (Allam 2018b, c). There is thus a highly competitive market where companies compete to tap into this profitable market without needing to drive the technologies further into these broader issues. Perhaps the drop in the branding of Smart Cities could be an opportunity to do what they always should have been doing – helping the world's cities become more inclusive, sustainable and climate resilient.

In an analysis carried out by Navigant (2018), the two main leaders in Smart City work are Cisco and Siemens which are closely followed by a set of contenders including IBM, Hitachi, Microsoft, GE, Schneider Electric and Bosch, among others (Fig. 1.4) (Navigant 2018). Sadowski (2016) also warns about the potential agenda of Smart City corporations in supporting a stand-alone profit-making agenda through the implementation of Smart City solutions. These commentators suggest that if cities invest in these corporations as part of their branding exercise rather than investing based on the values and visions derived from participatory approaches to governance as outlined by Nam and Pardo (2011), then smart technology may simply be wasted investment.

The attractiveness of the Smart City concept that seems to make large corporations seek to be associated with it is its market capitalization which has been rising as more cities were adopting the concept and its technologies. This has been

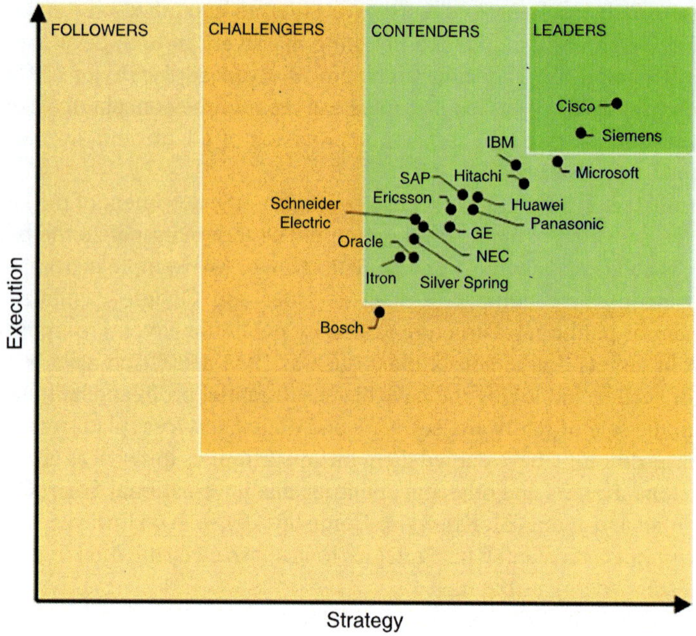

Fig. 1.4 Leader board of Smart City suppliers. (Navigant 2018; IMARC 2021)

expected to continue increasing, and it has been explained that, by 2025, the Smart City global market value will exceed its current value of $882.3 billion to above $2.5 trillion by 2027. In Europe alone, the report records that this concept will have the capacity to generate an annual revenue of over $120 billion from the current $50 billion. The Asian region will be the most active region in terms of implementation of the Smart City concept and is expected to drive substantial revenue generation.

The impressive revenue flow from Smart City projects is however seen, in the short term, to profit mainly corporations rather than cities. The reason is that the amount of investment required is deemed substantial and on the other side the return of investment is lengthy if linear business reporting is exercised. For instance, a report by Fishman and Flynn (2018) highlights that only approximately 16% of Smart City projects are self-funded, highlighting that a majority of the remaining percentage are funded via public-private partnerships (PPPs). Though this is among the potent alternatives to the financing of Smart City infrastructures and other bigger projects with wider value, numerous tax incentives are often further sought to encourage investment (Allam et al. 2018; Allam and Newman 2018a). A majority of Smart City initiatives are not meant for direct revenue generation but are meant to boost systems and operations that would in turn promote bigger goals of economic growth, security, resilience and sustainability. It is these bigger goals that draw funding and support from the public sector in cities and other levels of government to create these PPPs. The broader goals provide a conducive environment for associated revenue generating programs to be initiated. But do they actually deliver on those broader goals?

1.4 The Market Monopoly of Smart City Technology

A substantial number of large corporations have been positioning themselves for economic gains in the Smart City implementation arena. The race to gain competitive advantage in the provision of services, products, solutions and partnerships is apparent as depicted in Fig. 1.2. The competition is driven by the fact that numerous Smart City products like software, hardware, systems and expertise are required, but a majority of cities are not in a position to develop, install and maintain such products. Therefore, they contract and procure the services of corporations with capacity, both in terms of finances, skilled manpower and resources to install them on their behalf. As noted above, most Smart City projects are delivered by the adoption of PPP financing models; therefore, the winning corporation, or consortium, may have to incur almost all the cost, but in the long run, the endeavour becomes lucrative since they remain as the provider for the services, provision and maintenance of products and networks. This is even after the duration of contracts as most cities do not have the capacity for data gathering and analytics, installation and repair of smart components and/or running and maintaining systems (Calzada 2018).

This explains why large corporations invest heavily in research and development (R&D), developing patents and branding propriety technologies. This competition can be however detrimental to progress on broader goals for cities. As Yigitcanlar et al. (2018) decry, the core reason that prevents the actualization of a city's broader goals through the Smart City concept is the lack of standardisation of protocols and standards such that different Smart City components can interact with each other seamlessly. This confusion of technologies is not unusual in the history of cities as particular firms try to emphasise their specific capabilities rather than emphasise their value in integrated problem-solving. By adopting an isolated networking architecture, protocol and system, it means that small and local companies are shut out of the Smart City process. The change to a more inclusive and integrated set of technologies could easily open the Smart City system at any point to small and local firms. However, this would disrupt the market share and control of the large corporations. Jawhar et al. (2018) affirm this argument by showing how difficult it is to integrate services in Smart Cities for easy coordination and control. They argue that currently, most Smart Cities consist of such technologies like cyber physical systems (CPS), IoT, wireless sensors networks (WSNs) and cloud computing which are controlled by different corporations from a long distance away.

Zhang et al. (2016) delved into finding solutions for how the standardisation could be done and even proposed a framework that could be adopted to achieve collaboration, but the adoption of such options remains extremely untenable. Espada et al. (2019) also expressed the need for collaboration to these ends, but it is evident that this has not been achieved with competition being the reason behind the slow pace in finding a uniform protocol for similar services. Another argument is that with the heavy investment in R&D, large corporations are able to provide services and products at competing prices since their production and maintenance cost are greatly reduced. For smaller companies, offering such services or products at competitive prices would be difficult since in most cases, their operational costs are relatively higher.

Besides the above difficulties of harmonizing networking architecture and protocols, large corporations are also engaged in aggressive marketing campaigns aimed at promoting the concept of Smart City. Hollands (2015) outlines the motive behind this is profit-making, especially noting that ICT and digital connectivity are the main drivers of Smart Cities. He explains that by doing this, a sizeable number of cities fall prey to such corporations rather than utilising small and local companies, as large companies align themselves with the demands for development and implementation of the concept with quick turnaround times, even though the solutions are not contextualised. Similarly, Smart City campaigns are designed to evoke the sense of being left behind among cities that are not yet ‘smart’, and since the implementation of these projects is expensive, cities end up partnering with the large companies so controlling the entire project from design, development and implementation, operations and maintenance. By so doing, local companies and start-ups cannot match the technical and financial muscle of their larger competitors. This also can lead to issues of intellectual property of data and privacy concerns (Zoonen 2016).

Van Winden and van den Buuse (2017) acknowledge that large organisations are able to engage in larger projects since they usually have pilot programs which can be replicated in other localities when required. How the gap in market monopoly can be bridged is an interesting and complex issue that is explored later on in this book. An obvious intervention is the provision of incentives and support especially in pilot Smart City projects. This will provide small companies with financial capacity and exposure to train their staff and gain confidence to participate in larger projects. The incentives could be in the form of financial support, especially providing opportunities for the companies to secure loans and other support at reasonable rates, offering tax holidays, exemptions and the provision of domestic company protection mechanisms (Allam 2018a; Allam and Newman 2018a). Since most small projects and pilot projects are run by local municipalities and local players, Söderström et al. (2014) explain that small companies could gain experience by tapping into implementation opportunities, hence gaining confidence to transfer knowledge to larger projects when called upon.

1.5 Contextualising Smart City Technology

Van Winden and van den Buuse (2017) have demonstrated that the success of large corporations in implementing Smart City projects can be credited to their ability to perform numerous pilot projects and engage in intensive R&D on the best practices. They are able to offer technological solutions in the form of ‘off-the-shelf’ products that are easy and quick to implement. Nevertheless, evidence have shown such approaches of sourcing for 'off-the-shelf' solutions are not always the best, especially in providing optimal performance in complex and unique environments. The solutions are packaged in such a way that they seem to be able to address an array of issues that are common in most cities but often they miss the subtle differences of culture and place. Caird and Hallett (2018) explain that though there has been an increase in activities by the International Organization for Standardization (ISO), aimed at ensuring availability of standardised Smart City measurements, Smart City solutions are being promoted without showing that in fact they are not always compatible, scalable and replicable in a way that local areas can manage. For instance, Allam (2020) highlight that, in the case of privacy and security of Smart Cities, though there is a standardisation of techniques such as encryption, authentication and anonymity standards, among others, hackers from different regions find ways to navigate past these measures and compromise the security of systems and networks.

Bosch et al. (2017) suggest that there are many salient and unique issues synonymous with each city that render the cloning of such technological solutions as impractical. Chamoso et al. (2018) add that the customisation of technologies is not always easy since a majority of them do not provide high-level services that would allow developers to encapsulate the local needs coupled with cultural requirements. McKinsey and Company (2018a, b) further believe that off-the-shelf technological solutions are in most cases developed with the exclusion of stakeholders like sociologists, urbanists and other experts that have a key understanding of issues that confront cities and city fabrics. In addition, Bosch et al. (2017) argue that urban dwellers, who are the ultimate consumers of the Smart City technologies, are also not factored in, and this results in a bigger challenge once the project is implemented using such technologies. The result is a growing lack of acceptance of the concept from urban dwellers other than elites though these also can be bypassed as the Toronto example suggests, as set out in the Preface. Hamilton and Zhu (2018) argue that as issues of privacy and security emerge, citizens become reluctant to contribute to data sharing or in decision-making that relates to the city and how it ought to be managed. Berntzen et al. (2016) suggest that this trend results in difficult collaboration between policy makers, urban dwellers and other stakeholders in issues like smart waste management, optimal use of resources and adoption of other sustainability practices aimed at making the Smart City concept a reality in ways that benefit everyone (Newman 2020).

Costs associated with addressing localised issues related to privacy and security are noted to be expensive but necessary. The rise in cost is associated with the fact that it is not easy to customise 'smart' technologies to fit local needs, hence ensuring the role of technology developers long after a project's life cycle. Where customisation is possible, it has been argued that it becomes even more expensive as tailoring an off-the-shelf solution represents a considerable cost (Chêne 2009). On the other side, if the technological software is customisable with harmonised protocols in place, the operation and maintenance services could be left in the hands of small and local companies that would use local talent and resources, resulting in more competitive fees than large corporations. With an off-the-shelf solution, it may sometimes become hard to find a financier (Julian et al. 2015), especially in the case of PPP financial models as most financiers and developers would prefer to support a project that addresses the societal and cultural challenges in particular places, especially in cities, so as to ensure a direct influence and control.

To remain true to the purpose of adopting the Smart City technologies, Smart City solutions must be tailored and contextualised. This entails the engagement of various stakeholders when formulating policies, guidelines, designs and implementation strategies of the Smart City concept (Tomson 2017). Pereira et al. (2017) support that this is bound to allow for the seamless sharing of information enabled by increased collaboration between the municipalities, higher levels of government and engagement of community and other stakeholders. Through collaboration, the Smart City model and resulting technology can be adopted to provide customised local solutions. This can boost stakeholders' participation and confidence which is an essential ingredient in the success of a Smart City project. With tailored and contextualised solutions, local and small companies that understand the local challenges have the potential to participate in the implementation of projects, by being directly involved or in offering technical support. Similarly, urban dwellers would get the opportunity to secure employment opportunities as local businesses grow and new ones focusing on regeneration of all elements of the city emerge.

1.6 Redefining Smart Cities

The Smart City concept has demonstrated that it is a useful driver of urban change but that it lacks local contextualisation. Indeed, the concept of Smart City has largely created a fog about the issues of sustainability and climate resilience because it has suggested that these will somehow be addressed through the acquisition of digital data capacity. As will be shown in later chapters, it has probably been the most significant paradigm for much urban activity that has emerged out of the knowledge economy and its associated digital technology. But it is not a paradigm that can truly stand by itself as it is lacking inherent values for environmental, social and economic gain. 'Smart for what?' is constantly being raised now.

The illustration of the Toronto, a waterfront project quayside in the Preface, is highly symbolic on why Smart City is declining as a driving idea for cities. This is especially now that the post-COVID-19 economic agenda for cities needs financing and the Climate 100+ agenda is net zero with associated support of SDGs.

The key ideas in this next agenda of sustainability, climate resilience and inclusive economic development need all the help they can get, and this must include Smart City systems. The core of these net zero technologies are solar, batteries and electric vehicles, as well as the associated technologies of circular economy, resource efficiency and biophilic urbanism. These are best delivered at small-scale, local levels (Newman 2020; Green and Newman 2022). Thus, rebuilding the new economy will need a different economic model that makes the best of these technologies through much more local and contextualised development. This is much more able to then integrate the broader UN SDGs along with the net zero targets.

1.6.1 So Where Does Smart Cities Sit in This Emerging Economy?

In order to achieve these goals, there will need to be a growing and very significant utilisation of Smart City technology. The difference is that these technologies must be integrated into the broader environmental, social and climate values that must now be driving the development process. This must occur from the inception of projects whether they be from the Smart City private companies or the increasingly active public and private projects set up for net zero outcomes. Both of these systems can no longer be left in a silo that promises much but cannot deliver by itself. They need each other especially in the present dispensation where the impacts of COVID-19 coupled with the ongoing conflicts between Russia and Ukraine might prompt an increase in unsustainable practices. As such, it would be critical to redefine the Smart City programs to ensure they are not anchored, or dependent, on nonrenewable practices such as use of nonrenewable energies, as countries seek to stabilise their economies as well as shift their attention from traditional supply chains.

1.7 Conclusion

The Smart City is seen as being driven by ICT corporations looking to engage in a profit-making industry. The lack of harmonisation between the Smart City proponents has not only been making it harder for smaller, local companies to tap into established networks, but the competitive nature of the process has led to market monopolies that are now seen as driving the Smart City agenda, rather than any need to solve local and global problems. This can lead to economic disparities that can impact negatively on the urban social fabric. The need for contextualised technologies that feed into and drive the Smart City models is thus seen as paramount to ensure increased sustainability, resilience, economic development and inclusive dimensions. The reaction may be setting in as demonstrated in Toronto that we must 'kill' the Smart City. Or perhaps we can redesign the Smart City to enable it to find the deeper values of regenerative development that are emerging as the guiding next generation of urban planning. The other chapters will help explain this further.

References

Albino V, Berardi U, Dangelico RM (2015) Smart cities: definitions, dimensions, performance, and initiatives. J Urban Technol 22(1):3–21. https://doi.org/10.1080/10630732.2014.942092

Allam Z (2017) Building a conceptual framework for smarting an existing city in Mauritius: the case of Port Louis. J Biourban 4(1&2):103–121

Allam MZ (2018a) Redefining the smart city: culture, metabolism and governance. Case study of Port Louis, Mauritius. Curtin University, Perth, Australia

Allam Z (2018b) Contextualising the smart city for sustainability and inclusivity. New Design Ideas 2(2):124–127

Allam Z (2018c) On smart contracts and organisational performance: a review of smart contracts through the blockchain technology. Rev Econ Bus Stud 11(2):137–156

Allam Z (2020) Urban governance and smart city planning: lessons from Singapore. Emerald Group Publishing

Allam Z, Dhunny ZA (2019) On big data, artificial intelligence and smart cities. Cities 89:80–91. https://doi.org/10.1016/j.cities.2019.01.032

Allam Z, Jones D (2018a) Promoting resilience, liveability and sustainability through landscape architectural design: a conceptual framework for Port Louis, Mauritius; a small island developing state. Paper presented at the IFLA World Congress Singapore 2018, Singapore

Allam Z, Jones D (2018b) Towards a circular economy: a case study of waste conversion into housing units in Cotonou, Benin. Urban Sci 2(4):118

Allam Z, Newman P (2018a) Economically incentivising smart urban regeneration. Case study of Port Louis, Mauritius. Smart Cities 1(1):53–74

Allam Z, Newman P (2018b) Redefining the smart city: culture, metabolism & governance. Smart Cities 1:4–25

Allam Z, Dhunny A, Siew G, Jones D (2018) Towards smart urban regeneration: findings of an urban footprint survey in Port Louis, Mauritius. Smart Cities 1(1):121–133

Alomair B, Poovendran R (2014) Efficient authentication for mobile and pervasive computing. IEEE Trans Mob Comput 13(3):469–481. https://doi.org/10.1109/TMC.2012.252

Anthopoulos LG (2017) Understanding smart cities: a tool for smart government or an industrial trick? vol 22. Public administration and information technology, vol 1. Springer

Anttiroiko AV, Valkama P, Bailey SJ (2014) Smart cities in the new service economy: building platforms for smart services. Artif Intell Soc 29:323–334

Barkham R, Bokhari S, Saiz A (2018) Urban big data: city management and real estate markets. MIT Center for Real Estate and DUSP

Berntzen L, Johannessen MR, Florea A (2016) Smart cities: challenges and a sensor-based solution. Int J Adv Intell Syst 9(3 & 4):579–588

Berrone P, Ricart JE (2018) IESE cities in motion index. IESE Business School, University of Navara

Bezgrebelna M, McKenzie K, Wells S et al (2021) Climate change, weather, housing precarity, and homelessness: a systematic review of reviews. Int J Environ Res Public Health 18(11):5812

Bibri SE (2018) The IoT for smart sustainable cities of the future: an analytical framework for sensor-based big data applications for environmental sustainability. Sustain Cities Soc 38:230–253. https://doi.org/10.1016/j.scs.2017.12.034

Bosch P, Jongeneel S, Rovers V, Neumann H-M, Airaksinen M, Huovila A (2017) City keys indicators for smart city projects and smart cities. Citykeys

Braun T, Banjamin CMF, Iqbal F, Shah B (2018) Security and privacy challenges in smart cities. Sustain Cities Soc 39:499–507

Bruneo D, Distefano S, Giacobbe M, Minnolo AL, Longoa F, Merlino G, Mulfari D, Panarello A, Patanè G, Puliafito A, Puliafito C, Tapas N (2019) An IoT service ecosystem for smart cities: the #SmartMe project. Internet of Things 5(2019):12–33. https://doi.org/10.1016/j.iot.2018.11.004

Caird SP, Hallett SH (2018) Towards evaluation design for smart city development. J Urban Des:1–22. https://doi.org/10.1080/13574809.2018.1469402

Calzada I (2018) (Smart) citizens from data providers to decision-makers? The case study of Barcelona. Sustain For 10:3252. https://doi.org/10.3390/su10093252

Chamoso P, Gonzalez-Briones A, Rodríguez S, Corchado JM (2018) Tendencies of technologies and platforms in smart cities: a state-of-the-art review. Wirel Commun Mob Comput 3086854:1–17

Chêne M (2009) The implementation of integrated financial information management systems (IFMIS). Transparency International

Chourabi H, Nam T, Walker S, Gil-Garcia JR, Mellouli S, Nahon K, Pardo TA, Scholl HJ (2012) Understanding smart cities: an integrative framework. In: 2012 45th Hawaii international conference on system sciences, 4–7 January, 2012, pp 2289–2297. https://doi.org/10.1109/HICSS.2012.615

Cocchia A (2014) Smart and digital city: a systematic literature review. In: Dameri R, Rosenthal-Sabroux C (eds) Smart city. Progress in IS. Springer, Cham. https://doi.org/10.1007/978-3-319-06160-3_2

Dameri RP (2012) Searching for smart city definition: a comprehensive proposal. Int J Comput Technol 11(5):2544–2551. https://doi.org/10.24297/ijct.v11i5.1142

Datta A (2015) New urban utopias of postcolonial India: 'entrepreneurial urbanization' in Dholera smart city, Gujarat. Dialogues Hum Geogr 5(1):3–22

Espada JP, Yager R, Yu Z (2019) Communications, collaborations and services in networks of embedded devices. Futur Gener Comput Syst 92:560–563. https://doi.org/10.1016/j.future.2018.11.017

Fishman TD, Flynn M (2018) Using public-private partnerships to advance smart cities. Deloitte

Gascó-Hernandez M (2018) Building a smart city: lessons from Barcelona. Commun ACM 61(4):50–57. https://doi.org/10.1145/3117800

Giffinger R, Fertner C, Kramar H, Kalasek R, Pichler-Milanović N, Meijers E (2007) Smart cities: ranking of European medium-sized cities. Center of Regional Science (SRF), Vienna

Glasmeier AK, Nebiolo M (2016) Thinking about smart cities: the travels of a policy idea that promises a great deal, but so far has delivered modest results. Sustainability 8(11):1122

Google (2018a) Google trends compare. Google

Google (2018b) Google trends explore. Google

Graham S, Aurigi A (1997) Urbanising cyberspace? City 2(7):18–39

Green J, Newman P (2022) Transactive electricity markets: transactive electricity: how decentralized renewable power can create security, resilience and decarbonization. In: Droege P (ed) Intelligent environments 2 – advanced systems for a healthy planet. Elsevier

Hall RE, Bowerman B, Braverman J, Taylor J, Todosow H, Von Wimmersperg U (2000) The vision of a smart city. Brookhaven National Lab, Upton

Hamilton S, Zhu X (2018) Funding and financing smart cities. Deloitte

Harrison C, Eckman B, Hamilton R, Hartswick P, Kalagnanam J, Paraszczak J, Williams P (2010) Foundations for smarter cities. IBM J Res Dev 54(4):1–16. https://doi.org/10.1147/jrd.2010.2048257

Hollands RG (2008) Will the real smart city please stand up? City 12(3):303–320. https://doi.org/10.1080/13604810802479126

Hollands RG (2015) Critical interventions into the corporate smart city. Camb J Reg Econ Soc 8(1):61–77. https://doi.org/10.1093/cjres/rsu011

Huertas A, Moreno A, Pascual J (2021) Place branding for smart cities and smart tourism destinations: do they communicate their smartness? Sustainability 13(19):10953

Huiling E, Goh B (2017) AI, robotics and mobility as a service: the case of Singapore. Field Actions Sci Rep Special Issue 17:26–29

IBM (2010) Smarter thinking for a smarter planet. IBM

IMARC (2021) Smart cities market: global industry trends, share, size, growth, opportunity and forecast 2022-2027. Available at imarcgroup.com/smart-city-market. Accessed 20 June 2022

Ishida T (2017) Digital city, smart city and beyond. In: The 26th world wide web international conference (WWW17), Perth, Australia, pp 1151–1152

Jawhar I, Mohamed N, Al-Jaroodi J (2018) Networking architectures and protocols for smart city systems. J Internet Serv Appl 9(26). https://doi.org/10.1186/s13174-018-0097-0

Joss S, Sengers F, Schraven D, Caprotti F, Dayot Y (2019) The smart city as global discourse: storylines and critical junctures across 27 cities. J Urban Technol 26(1):3–34. https://doi.org/10.1080/10630732.2018.1558387

Julian C, Lucy T, Farr J (2015) Commercial-off-the-shelf selection process. Eng Manag J 31(5):63–71. https://doi.org/10.1080/10429247.2011.11431896

Khan M, Babar M, Ahmed SH, Shad SC, Han K (2017) Smart city designing and planning based on big data analytics. Sustain Cities Soc 35:271–279. https://doi.org/10.1016/j.scs.2017.07.012

Kitchin R (2014) Making sense of smart cities: addressing present shortcomings. Camb J Reg Econ Soc 8(1):131–136. https://doi.org/10.1093/cjres/rsu027

Kolotouchkina O, Seisdedos G (2017) Place branding strategies in the context of new smart cities: Songdo IBD, Masdar and Skolkovo. Place Brand Public Dipl 14(2):115–124. https://doi.org/10.1057/s41254-017-0078-2

Mancebo F (2020) Smart city strategies: time to involve people. Comparing Amsterdam, Barcelona and Paris. J Urban Int Res Placemaking Urban Sustain 13(2):133–152

Mastercard (2017) Defining what makes a city a destination. https://newsroom.mastercard.com/press-releases/defining-what-makes-a-city-a-destination/. Accessed 4 Mar 2019

McKinsey & Company (2018a) Smart city solutions: what drives citizen adoption around the globe? Mckinsey Center for Government

McKinsey & Company (2018b) Smart cities: digital solutions for a more livable future. McKinsey & Company, Brussel

Mordor Intelligence (2018) Smart cities market size, share – segmented by solution (smart mobility management, smart public safety, smart healthcare, smart buildings), and geography – growth, trends and forecast (2019–2024). https://www.mordorintelligence.com/industry-reports/smart-cities-market. Accessed 4 Mar 2019

Mosenia A, Jha NK (2017) A comprehensive study of security of internet of things. IEEE Trans Emerg Top Comput 5(4):586–602. https://doi.org/10.1109/TETC.2016.2606384

Nam T, Pardo TA (2011) Conceptualizing smart city with dimensions of technology, people, and institutions. In: Proceedings of the 12th annual international digital government research conference: digital government innovation in challenging times. ACM, pp 282–291

Natural Resources Defense Council (2018) What are smarter cities? http://smartercities.nrdc.org/about. Accessed 13 May 2018

Navigant (2018) Navigant research leaderboard: smart city suppliers. https://www.navigantresearch.com/research/navigant-research-leaderboard-smart-city-suppliers. Accessed 3 June 2018

Newman P (2020) COVID-19, cities and climate: historical precedents and expected transitions for the new economy. Urban Sci 4:32. https://doi.org/10.3390/urbansci4030032

Northstream (2010) White paper on revenue opportunities. Northstream, Stockholm

Pereira GV, Cunha MA, Lampoltshammer TJ, Parycek P, Testa MG (2017) Increasing collaboration and participation in smart city governance: a cross-case analysis of smart city initiatives. J Inf Technol Dev 23(3):526–553. https://doi.org/10.1080/02681102.2017.1353946

Richter C, Kraus S, Syrjä P (2015) The smart city as an opportunity for entrepreneurship. Int J Entrep Ventur 7(3):211–226

Rohaidi N (2018) Singapore wins smart city of 2018 award. GI. https://govinsider.asia/smart-gov/singapore-wins-smart-city-of-2018-award/. Accessed 5 Mar 2019

Sadowski J (2016) Selling smartness-visions and politics of the smart city. Arizona State University

Satterthwaite D (1997) Sustainable cities or cities that contribute to sustainable development? Urban Stud 34(10):1667–1691. https://doi.org/10.1080/0042098975394

Sepasgozar SME, Hawken S, Sargolzaei S, Foroozanfa M (2018) Implementing citizen centric technology in developing smart cities: a model for predicting the acceptance of urban technologies. Technol Forecast Soc Change (In Press)

Söderström O, Paasche T, Klauser F (2014) Smart cities as corporate storytelling. City 18(3):307–320

Sofeska E (2017) Understanding the livability in a city through smart solutions and urban planning toward developing sustainable livable future of the city of Skopje. Procedia Environ Sci 37:442–453

Souza A, Figueredo M, Cacho N, Araújo D, Prolo CA (2016) Using big data and real-time analytics to support smart city initiatives. IFAC-Pap Online 49(30):257–263

Su K, Li J, Fu H (2011) Smart city and the applications. Paper presented at the 2011 International Conference on Electronics, Communications and Control (ICECC), 2011/09

Taylor S (2014) How service providers can profit from smart cities. https://blogs.cisco.com/sp/how-service-providers-can-profit-from-smart-cities. Accessed 4 Mar 2019

To W-M, Lee PKC, Lam K-H (2018) Building professionals' intention to use smart and sustainable building technologies – an empirical study. PLoS One 13(8):e0201625. https://doi.org/10.1371/journal.pone.0201625

Tomson T (2017) Understanding the contextual development of smart city initiatives: a pragmatist methodology. She Ji: J Design Econ Innov 3(3):210–228. https://doi.org/10.1016/j.sheji.2017.11.004

Toppeta D (2010) The smart city vision: how innovation and ICT can build smart, "livable", sustainable cities. Innov Knowl Found 5:1–9

Trends G (2018) Smart cities popularity. https://trends.google.com/trends/explore?q=smart%20cities. Accessed 25 May 2018

Trindada EP, Hinning MPF, Costa EM, Marques JS, Bastos RC, Yigitcanlar T (2017) Sustainable development of smart cities: a systematic review of the literature. J Open Innov Technol Mark Complex 3(11). https://doi.org/10.1186/s40852-017-0063-2

United Nations Smart cities and infrastructure: Report of the Secretary-General. In: Economic and Social Council (ed) Commission on science and technology for development, Geneva, 9–13 May 2016, pp 1–18

Vale LJ (2014) The politics of resilient cities: whose resilience and whose city? Build Res Inf 42(2):191–201. https://doi.org/10.1080/09613218.2014.850602

van den Besselaar P, Beckers D (1998) Demographics and sociographics of the digital city. In: Ishida T (ed) Community computing & support systems, Lecture notes in computer science, vol 1519. Springer, Berlin, pp 108–124

Van Winden W, van den Buuse D (2017) Smart city pilot projects: exploring the dimensions and conditions of scaling up. J Urban Technol 24(4):51–72. https://doi.org/10.1080/10630732.2017.1348884

Washburn D, Sindhu U, Balaouras S, Dines RA, Hayes N, Nelson LE (2009) Helping CIOs understand "smart city" initiatives. Growth 17(2):1–17

Wilhelmsson M, Ismail M, Warsame A (2021) Gentrification effects on housing prices in neighbouring areas. Int J Hous Mark Anal (ahead-of-print)

Yigitcanlar T, Kamruzzaman M, Buys L, Ioppolo G, Sabatini-Marques J, da Costa EM, Yun JJ (2018) Understanding "smart cities": intertwining development drivers with desired outcomes in a multidimensional framework. Cities 81:145–160. https://doi.org/10.1016/j.cities.2018.04.003

Zhang N, Chen J, Chen X, Chen J (2016) Semantic framework of internet of things for smart cities: case studies. Sensors 16(9):1501. https://doi.org/10.3390/2Fs16091501

Zoonen LV (2016) Privacy concerns in smart cities. Gov Inf Q 33(3):472–480. https://doi.org/10.1016/j.giq.2016.06.004

Chapter 2
Smart Cities and Sustainability: How Smart City Helps with Sustainability

Abstract The challenges and impacts from rapid urbanisation coupled with the impacts of climate change and other global planetary boundary issues are prompting cities to take urgent action toward safeguarding the sustainability of the urban fabric – reducing environmental and social impacts while improving liveability. The advent of sustainability-oriented technology is being recognised as having a predominant role in this process. However, these solutions are often claimed as part of the Smart City technology arsenal when often they have little to do with digital data systems. Thus, the agenda of Smart Cities in the past has claimed digital technology upgrades will automatically help solve sustainability problems; however, the simple provision of more digital capacity does not necessarily mean this will happen. New approaches to sustainability where smart systems are made an integrated part of the metabolism of cities can provide solutions that also can lead to increased liveability levels.

2.1 Introduction

There are nine planetary boundaries identified that nations and cities need to address (Steffen et al. 2015). Cities are able to address most of these through their energy, water, transport, industry and waste systems with various scales of action (Meyer and Newman 2020). Perhaps the biggest of the boundaries and certainly the one attracting the most attention in the 2020s is climate change.

The impacts of climate change are now urgent as highlighted by the recent report of the IPCC (IPCC 2018), and the role of cities in this process is apparent (International Energy Agency 2017a, b). The IPCC reports highlight that a global warming of 1.5 °C above preindustrial levels would have devastating consequences on the entire global ecosystem and human activities are responsible for an increase in temperatures by between 0.8 and 1.2 °C. If mitigation measures are not hastily implemented, by 2030 and beyond, the rise would increase from 1.5 to 2 °C and beyond (almost 3.5 °C) (IPCC 2022), and the consequences will be dire. With the rise of 0.8 °C, it is reported that the global sea level has risen by approximately 1.7 cm, and this would continue to increase in tandem to increase in temperatures.

Z. Allam, P. Newman, *Revising Smart Cities with Regenerative Design*, Cities and Nature, https://doi.org/10.1007/978-3-031-28028-3_2

The World Bank (2010) projected that such increases in temperature would lead to a permanent GDP reduction of between 4% and 5% in South Asian and African countries.

The need to address climate change is addressed in more detail in the next chapter. Here we want to show how cities need to be understood as systems that not only create wealth and liveability, but they are like natural system metabolism, converting resources like energy, water and materials into urban fabric but at the same time inevitably creating waste that must be dealt with; otherwise, it has global and local impacts. Sustainability in cities can be framed as 'reducing the resources, reducing the wastes and increasing the liveability at the same time' (Newman and Kenworthy 1999b). In this chapter, we will examine the growing lists of issues on the agenda of sustainability in the local and global arena, how this can be applied to cities using the metabolism approach and how smart systems can be used to assist in delivering more sustainable cities.

2.2 Sustainability Issues in Cities

Cities, as is any given economy, are the engines of economic growth (Collier et al. 2018). Mitra and Mehta (2011) and the World Bank (2010) show that large percentages of national GDPs are attributable to cities. To achieve this, cities consume a large amount of resources – water, minerals, energy, food, land and forests, among many others. Bergesen et al. (2017) contend that in the process of availing these resources, large amounts of pollutants and greenhouse gases are emitted which have potential to affect global and local environments such as the increase in global temperatures (IEA and UNEP (2018) report). Among the leading contributors of these pollutants include traditional power plants relying on fossil fuels (Narayanan et al. 2019), automobiles (Newman and Kenworthy 1999a), factories emitting poisonous gases and liquid effluents that find their ways in the soils and water aquifers (Tripathi 2017). In addition, the increase in global population allied with an equal increase in consumerism directly contributes to environmental impacts. Kalmykova et al. (2016) explain that twentieth-century housing, electronics and automobiles are products that are resource intensive and are mostly sourced from nonrenewable sources. Similarly, when most of these products complete their life cycle, they are disposed, rather than recycled, leading to increased levels of pollution and subsequent environmental degradation.

The unprecedented rates of urbanisation and the rapid increase in urban population which Kaneda et al. (2020) predict will reach over 68% by 2050 are among the factors which contribute to the overexploitation of natural resources and pose threats to human survival. With such trends, it became urgent from the 1970s on that many changes needed to be made and hence the notion of sustainable development was created through the UN's process called the *World Commission on Environment and Development* (WCED 1988).

Cities were soon made part of this sustainability agenda, and a series of approaches were created for how urban planning could become part of the solution, not the problem: sustainable city (Newman and Kenworthy 1999b), eco-city (Cugurullo 2017), resilient city (Stumpp 2013) and low carbon city (Shen et al. 2018) were adopted at different periods. They are all geared toward the application of knowledge to urban planning with a shared goal of increasing sustainability in various ways.

Slowly the experimentation using these different models has yielded tangible results which include using environmentally friendly construction materials, promoting the importance of mixed land use like in the case of Singapore (Ong 2017) and emphasising the need for reduction in use of automobiles in favour of cycling and using transit initiatives like is advanced in the 15-Minute City concept (Allam et al. 2018a, b). The application of digital technology could have been immediately applied to accelerating the sustainability solutions using these models. But they were not.

2.3 The Smart City Takeover

According to Van Winden and van den Buuse (2017), all the ideas brought forward in the above models have converged to a new urban planning concept known as Smart City. The Smart City agenda was born out of the global financial crash when the digital economy was created using digital technology. This concept was hailed for its unique economic benefits that could be brought about by its ability to spur efficiency and increase performance in the social, economic, environmental and political spheres. For instance, the adoption of the Smart City concept was suggested to help in boosting the attractiveness of cities, thus boosting tourism activities, attracting foreign direct investments (FDIs) and, on the local front, promoting employment opportunities, among many other things (Anand and Navío-Marco 2018). It should be noted that the list did not refer to any sustainability outcomes but instead assumed that if the above outcomes were obtained, then the other sustainability goals would follow. They did not. In fact, the focus on Smart City projects had the unintended consequence of marginalizing important sustainability and climate change agendas, despite their success in attracting investment and promoting economic growth.

This fogging of the sustainability agenda by Smart City proponents can be seen by many other commentators who made the same error of assuming that the simple increase in digital capability would automatically solve environmental problems. Anthopoulos (2017) explains that the concept of Smart City, via a range of diverse technologies, has brought numerous smart innovations in urban centres that are proving positive in addressing the impacts of climate change – though these are rarely quantified by the author. Macháč et al. (2016) point at such trends like the adoption of green and blue infrastructure that help in reducing the intensity of extreme weather conditions as well as improving the quality of air that in most cases is polluted from different pollutants in the city. None of these technologies were digital smart systems; they were just simply direct sustainability-oriented technologies branded as smart.

In the same way, Blanco et al. (2018) argue that the technologies utilized in Smart City projects have allowed for smarter urban planning strategies allowing for pockets of green spaces in cities that serve as recreation parks and help the city dwellers to maintain a healthy lifestyle. These are quite simply normal urban planning that has been known and delivered for hundreds if not thousands of years. They do not need to be called smart. Blanco's suggestions are simply good planning with a series of sustainability benefits such as (a) protecting some urban biodiversity as well as reducing the amount of hard surface runoff, thus easing flooding in cities, and (b) allowing for the densification of urban areas and thus allowing for the optimisation of space, as well as resettlement of numerous households. There is no need to suggest these will automatically happen if digital data capacity is increased but perhaps it would have happened if the smart systems had been integrated into real projects delivering these broader goals of city management.

Smart cities were also claimed to be implicitly resource efficient by Kylili and Paris (2015) who hail the zero energy buildings that are deemed crucial in helping in the reduction of emissions by their ability to allow for 100% consumption of alternative, renewable energy. None of these approaches are explained in terms of the digital systems needed, but instead they are simply branded good sustainability practice as 'Smart City'. This is dangerous as the concepts led to many cities prioritizing and procuring digital capacity with a rationale that this would automatically create sustainability outcomes. They are not going to do that unless they are designed for that purpose.

Liveability outcomes were also claimed to be automatically linked to the Smart City. For example, Appio et al. (2019) suggest that Smart City concepts have the promise of increasing the liveability of cities while enabling the vision of cleaner environment to be achieved. Calvillo et al. (2016) highlight that this is possible through its emphasis on the use of cleaner energy in households and in other areas like street lightings, in industries and in transport and communication sectors, among others. However, this may well have simply been greenwashed as the majority of Smart City technology that was actually procured through this period was about surveillance, often rationalised as improved liveability. This could be affirmed by considering the case of China, which currently has the highest number of cities branded as 'smart', yet it is among the leading countries in terms of emission.

The liveability index of Smart Cities was suggested to help in improving the security in all aspects of the city. With technologies such as big data, artificial neural networks (ANNs) (Allam 2019), Internet of Things (IoT) and others, the numerous technological components installed in cities can allow for real-time gathering and analysis of data that is essential for immediate and contextualised solutions and responses to urban challenges. These could have been attached to sustainability outcomes but invariably were only about how to control human behaviours. Such technologies and components allow for the prediction and analysis of human behavioural trends (Mahdavinejad et al. 2018) and thus were rapidly picked up by companies wanting to manipulate consumption of anyone whose data they could obtain and even been used to manipulate election outcomes by shaping individualised messaging.

These same Smart City technologies could have been used for citizen participation in the planning and management of cities through data sharing via social media platforms and bridging the gap of social inequality (McKinsey and Company 2018; Martinez-Balleste et al. 2013). The same Smart City technology could have been used to achieve increased sustainable outcomes in every aspect of city life in both low-income and high-income cities. They could have been and should have been focused on sustainability, but were not.

2.4 Sustainability and The Metabolism of Cities

Cities can be viewed as living organisms and as entities that sustain life in an organic way, as portrayed by Jane Jacobs (1961), Christopher Alexander (2002) and Nikos Salingaros (2014). The analogy of a city as a living organism is particularly useful in helping to understand sustainability in an urban planning context. Urban metabolism is the study of input and output processes of cities to understand how the structural issues about sustainability depend on understanding the resources which are the core inputs, the wastes that are the related core outputs and the desired liveability which results from this metabolism. The goal of sustainability is to reduce the resources and reduce the wastes while simultaneously increasing the liveability (Newman 1999). See Fig. 2.1.

Proper understanding of the urban metabolism concept is an important step toward the pursuit of sustainability outcomes, as this will mean an optimal utilisation of available resources and better use of planning designs more responsive to modern sustainable agendas, as captured in the SDG11, the New Urban Agenda and

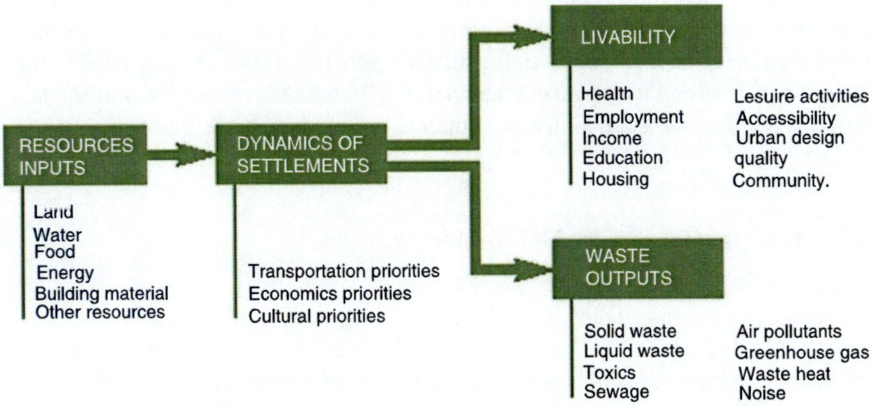

Fig. 2.1 Extended metabolism model of human settlements

other initiatives. The mechanisms for doing these changes are suggested in Fig. 2.1 to be cultural, economic and transportation priorities.

(a) Cultural changes are brought through the community. As Salingaros (2005) suggests, such an understanding would ensure that humans and their interactions with each other and with the urban fabric will be determining factors in urban planning. Salingaros (2006) believes that technologies especially those with the potential to allow for compact cities are the solution to addressing sprawl but this requires good community facilities to be part of the compact city. He argues that such technologies have the potential to impart life to more compact apartments and office towers that have for long been erected without considering their value in advancing sustainability. Similarly, Newman, Beatley and Boyer (2017) also advocate for inclusion of biophilic urbanism systems that would promote cyclical and regenerative metabolism and work best in compact cities.
(b) Economic priorities are about how best to provide infrastructure for all the cities' needs and to do this using economic incentives that reduce the need for resources and enable wastes to be recycled. Newman (2020a, b) argues that cities need to invest in renewable and distributed energy, create sustainable mobility systems and focus on ensuring inclusivity and healthy cities as part of the next economy.
(c) Transportation priorities create the different kinds of urban fabric including the transport infrastructure suggested above. The different parts of the city require different transport infrastructure solutions and highlight the value of local, contextual solutions.

Smart city technologies are not listed in these mechanisms, but in reality, the delivery of each of them outlined above is much easier and more productive if Smart City systems are integrated into each of them. For example, it is possible to apply real-time data generation and analytics and real-time monitoring of different aspects of automation to each of the cultural, economic and transportation systems outlined. If integrated in from the start, the Smart City can help make sustainability happen. It is possible to achieve wholeness, resiliency and sustainability in cities while also improving the liveability status, thus actualising the benefits of shaping cities using the concept of urban metabolism (Newman 1999), but the two big ideas of the last 30 years in cities – sustainability and Smart Cities – must be integrated.

2.5 Integrated Smart Urban Metabolism

Faced with increasing challenges tied to urbanisation and rapid population increase, cities are in dire need to achieve more sustainable levels of urban metabolism, and the role of technology is paramount in this process. Chávez et al. (2018) acknowledge this proposition and argue that technology is important in helping stakeholders understand, plan, implement and track urban metabolism especially if the future of the urban fabrics and components is to be secured. The need for technology is propagated by the fact that urban centres have been known to consume approximately 75% of the global natural resources (UN Habitats 2018) and, in return, generate over 50% of global solid wastes and between 60% and 80% of greenhouse gas emissions and other pollutants (UNEP 2016). IRENA (2018) records that over 85% of energy consumed in majority of cities is from nonrenewable energy sources like fossil fuels and only a meagre 15% is derived from renewable energy sources.

Such figures are expected to keep increasing, as it is projected that over 68% of the global population will be living in urban centres and these are all growing bigger, especially large cities. For instance, according to the United Nations (2018) in 2016, there were approximately 512 cities globally that hosted at least one million people. By 2030, these cities are projected to increase to approximately 662 cities. In addition, megacities that have more than ten million people are expected to increase from 31 cities, recorded in 2016, to about 41 cities by 2030 (United Nations 2016; UN Population Division 2019). Such numbers will mean an increase in consumption of different resources and, at the same time, an increase in waste generation. Nevertheless, with technologies such as Smart City technologies and systems integrated with sustainability-based technologies and systems, it is possible that smart and sustainable urban metabolism can be achieved.

Integrated approaches to smart and sustainable cities are being trialled (Allam and Jones 2021). For instance, in the energy sector, alternatives like photovoltaic-solar energy, hydro energy, wind energy and others are being rapidly introduced in all countries (Motyka et al. 2018; REN21 2018) but will only be able to truly take over from fossil fuels when smart systems are used to integrate them into grids both large and small, especially into the rapidly emerging distributed grids which require local integration (Green and Newman 2022).

Fan et al. (2019) express how the use of technology in managing urban fabric has had significant impacts on the energy-food and energy-water nexus, where the emphasis is on optimising sustainability approaches that integrate these different elements. Restrepo and Morales-Pinzón (2018) explain that by using modern technological advancement such as the use of big data, AI, IoT and machine learning, cities are now able to track, and project in real time, the inflow and outflow of materials; hence, informed and decisive actions are taken both in optimising resource use and also in management of wastes. Gaigné et al. (2012) highlight that by doing this, city managers are able to formulate the best approaches that promote growth and, at the same time, emphasise the sustainable agenda. But the initiative to integrate these elements is rarely part of the contracts with the responsible government agencies; instead, they are left with the branded package of smart systems and need to work out the next steps themselves.

Besides being focal in providing efficient resource management and increased performance, with reduced wastage of resources, technology can foster changes of behaviour in urban dwellers, who are frequently much concerned about their consumption behaviour (Samad et al. 2019; McKinsey and Company 2018). In particular, technologically inclined concepts such as the Smart City are said to accommodate and promote the participation of citizens in the governance of the cities, especially through information sharing. By so doing, Stewart et al. (2017) report that most urban dwellers where such technologies are fully integrated are said to embrace the need for resource optimisation, and a majority are actively participating in 'smart' initiatives like the use of bicycles and public transport systems instead of owning automobiles that contribute to harmful emissions. A case in point is the city of Copenhagen that is said to benefit from cycling culture of the locals where over 90,000 tonnes of CO2 that could have reached the environment are saved, thus improving the sustainability of the city (International Organization for Standardization 2015). However most of these initiatives for cycling and walking in Copenhagen were done well before any smart systems came along (Matan and Newman 2016).

Cities are also active in adopting what Newman and Thomson (2018) called 'cyclical use of resources' (recycling), thus reducing the amount of waste. For instance, Shahrokni et al. (2015) explain how the city of Stockholm has managed to optimise waste management and recycling. However, it is not possible to claim that this has been done by adopting a data centred approach. The data are now gathered from smart components installed in cities as well as from urban dwellers, but this builds on decades of non-smart initiatives (Newman and Kenworthy 1999b).

Smart technology can be integrated into urban metabolism to accelerate how most urban areas can achieve increased liveability dimensions by adopting better and more efficient management of resources. The emphasis on the adoption of mixed land use, smart waste management and adoption of cleaner renewable energies is possible but must be integrated into mainstream sustainability practice.

2.6 Infusing Smart Technology into Sustainability in Cities

Cities and urban centres are in a constant process of change. The industrial revolution prompted the transformation of most western cities from medieval to industrial cities (International Federation for Housing and Planning 2016). When automobiles came into existence in the twentieth century, there was a demand for further transmissions, and the resulting urban models led to an increase in central business districts characterised by high-rise buildings, improved transportation systems and also increased urban sprawl, as people could afford to travel to areas far from cities. In the recent past, the advent of ICT and its ability to seamlessly integrate into the urban fabric has seen a rise in new high-tech urban developments which have been actualised in various forms (Allam 2017). Information and communication systems based on digital data are the newest kind of technology in cities and these as cities seek to improve economic, social and environmental outcomes. Now cities are planning technoparks, technopoles, science parks and eco-cities which all feature these new smart technologies (Oh and Phillips 2014). The newest planning model that is particularly gaining substantial attention after the realities of the COVID-19 pandemic is the 15-Minute City model. The model proposes a redefinition of urban areas such that besides them being 'smart', they also incorporate aspects of proximity, accessibility, diversity and ubiquitousness (Allam et al. 2020). This way, some of the existing challenges in cities such as traffic, prompted by increased number of private cars, would be addressed. Further, this model proposes prioritisation of the human dimensions well defined in SDG11; hence, its successful adoption would not only allow cities to be only smart, but they would be vibrant with human oriented outcomes.

2.6.1 How Then Do We Infuse Smart City Technology into Urban Sustainability Programs?

The emphasis on the use of technologies in those concepts demands a restructuring of how we manage cities. The rest of the book will show this can be done starting with climate change and then with the regenerative design approach and finally how to build it into economic development as an inclusive process.

In all these chapters, it will be important to see how the theory of urban fabrics can be used to find local and community-based approaches, to regenerate each part of the city differently using integrated infrastructure. The theory of urban fabrics is outlined briefly using examples of how it can be applied to integrate the smart and sustainable/net zero technologies:

1. Old central cities have a fabric based on walking as that was how they were formed around the potential to walk to everything in half an hour, and hence all the core infrastructure needs to support walkability in these areas. Smart and sustainable infrastructures need to fit together to make walkability more attractive for everyone and not just safer. This means better lighting and shared transport systems as well as shared solar energy managed by shared data that enables a highly intensive knowledge economy to thrive.
2. Nineteenth- and early twentieth-century transit fabric is spread along corridors, and the medium density fabric was built around reaching all the key parts of the city in half an hour by train or tram. These areas need to regenerate the transit systems with electric vehicles of all kinds, especially new twenty-first-century e-rideables and trackless trams, buildings with mixed use and shared solar for all power and recharging of vehicles with special economic uses for health, education and shopping services.
3. Twentieth-century car-based suburbs that are designed around reaching everything by car in half an hour. The low-density suburbs are important for large warehousing and industry as well as housing but are increasingly creating small urban centres that enable transit and walkability for these outer areas. The opportunities for solar are more extended as well as other aspects of the circular economy such as waste recycling. Large precincts can integrate smart and sustainable infrastructure.
4. Peri-urban and rural villages are increasingly seeking ways to use smart and sustainable infrastructure that can enable a large degree of self-sufficiency with

access to the adjacent city when specialised services are needed. The need for community-based smart systems to enable complete net zero outcomes is feasible here.

The alternative approach taken by many nations is to establish new cities. There is not a great history in any of these new cities, so they are becoming rarer. Although new cities would be a breath of fresh air to the planner and some urban dwellers, building a city from scratch is not always a sustainable endeavour as it creates demand for more resources (Slavova and Okwechime 2016) and generally increases dependency on unsustainable practices like the use of automobiles. This is true since available lands for such projects in most countries are relatively far from existing cities; hence, people are forced to travel using fossil fuel powered vehicles, since public encouragement in the form of fiscal incentives for energy saving and new kinds of transit services is generally left out of these new cities. Shepard (2017) suggests that these challenges relate to the expensive nature and the amount of time required to consider them to make new cities ready and self-reliant. Tosics (2015) showed how residential units expected to accommodate over 20,000 people in a new town of Aspern Seestadt, Vienna, are projected to take over 20 years to complete, and it was started in 2015. He acknowledges, like Salingaros (2006), that such new projects are unsustainable and proposes the adoption of compact residential neighbourhoods in existing cities. We would suggest regenerating all parts of the city with new twenty-first-century infrastructure using the theory of urban fabrics is likely to be much more successful.

Economically, a brand-new city will pose an unfair advantage as businesses and most lucrative economic activities may be bound to flock to the new cities leaving existing ones in a slow state of urban decay. Inclusion is more likely if regeneration happens inside the present urban fabrics. However, they need to also be inclusive in the sense of retaining the qualities of place in each urban fabric. Jacobs (1961) outlined in her famous book *The Death and Life of Great American Cities* that existing cities were built around some indelible heritage that identify and relate with the identity of the place and a unique sense of belonging to locals. She was reacting to the freeways and high-rise public housing used to clear out older urban fabric in New York with a complete loss of places with significant heritage. These places lost all their economic activities as well because local people had fashioned their businesses to accommodate to the unique local urban fabric. The theory of urban fabrics starts by recognising the value of each place, respecting its functionality and then seeking to regenerate this by new, integrated twenty-first-century infrastructure.

2.7 Conclusion

While smart technology is a strong component of urban futures and its role in achieving urban regeneration is significant, the most important quality to be sought in any urban process is to ensure the deeper values of place are respected. There is a strong tradition in architecture and urban planning that outlines this approach through authors like Salingaros (2006, 2014), Alexander (1965, 1979, 2002) and Jacobs (1961). The contribution of this book is that such place values now need to include values about climate resilience, net zero and broader reductions in urban metabolism and liveability. These are defining the next economy and need to be attended to rapidly within our cities.

But these deeper values cannot be simply put aside by smart technology branding exercises that assert that all the big issues will be dealt with by some kind of magic. The reality is that some cities have been getting increased digital data capacity, but no other sustainability goals are achieved. Cities must be geared toward increasing the 'wholeness' of the city, and as we suggest, cities need to bring in smart technology as part of an integrated package that will simultaneously improve the urban metabolism as well as the liveability status.

Allam and Newman (2018) argue that the integration of technology in existing cities allows citizens to be participants in their management by sharing vital data that help improve areas like security but could also allow other resource and waste management objectives. Kraus et al. (2015) suggested that smart technologies can also allow locals to be more innovative and provide them with opportunities to exploit entrepreneurial opportunities that in turn lead to improved economic status and job creation and, in extension, help in reducing economic and social inequalities. This book suggests that it is no longer acceptable to simply say Smart City technologies 'allow' such fine objectives, but they must be simultaneously applied to create these outcomes, or else they are simply a branding exercise.

References

Alexander C (1965) A city is not a tree. Archit Forum 122(1):58–61

Alexander C (1979) The timeless way of building. Oxford University Press, New York

Alexander C (2002) The nature of order: the process of creating life. The Centre for Environmental Structure, Berkeley

Allam Z (2017) Building a conceptual framework for smarting an existing city in Mauritius: the case of Port Louis. J Biourban 4(1&2):103–121

Allam MZ (2018a) Redefining the smart city: culture, metabolism and governance. Case Study of Port Louis, Mauritius, vol 1. Curtin University, Perth, p 4

Allam Z (2018b) Contextualising the smart city for sustainability and inclusivity. New Des Ideas 2(2):124–127

Allam Z (2019) Achieving neuroplasticity in artificial neural networks through smart cities. Smart Cities 2(2):118–134

Allam Z, Jones DS (2021) Future (post-COVID) digital, smart and sustainable cities in the wake of 6G: digital twins, immersive realities and new urban economies. Land Use Policy 101:105201

Allam Z, Newman P (2018) Redefining the smart city: culture, metabolism and governance. Smart Cities 1(1):4–25

Allam Z, Moreno C, Chabaud D et al (2020) Proximity-based planning and the "15-minute city": a sustainable model for the city of the future. In: The Palgrave handbook of global sustainability. Springer International Publishing, Cham, pp 1–20

Anand PB, Navío-Marco J (2018) Governance and economics of smart cities: opportunities and challenges. Telecommun Policy 42:795. https://doi.org/10.1016/j.telpol.2018.10.001

Anthopoulos LG (2017) Understanding smart cities: a tool for smart government or an industrial trick? vol 22. Public Administration and Information Technology, Vol 1. Springer International Publishing, Cham

Appio FP, Lima M, Paroutis S (2019) Understanding smart cities: innovation ecosystems, technological advancements, and societal challenges. Technol Forecast Soc Chang 142:1–14. https://doi.org/10.1016/j.techfore.2018.12.018

Bergesen JD, Suh S, Baynes T, Musango JK (2017) Environmental and natural resource implications of sustainable urban infrastructure systems. Environ Res Lett 12(12):1–24. https://doi.org/10.1088/1748-9326/aa98ca

Blanco JL, Fuchs S, Parsons M, Ribeirinho MJ (2018) Artificial intelligence: construction technology's next frontier. Build Econ 2018:7–13

Calvillo CF, Sánchez-Miralles A, Villar J (2016) Energy management and planning in smart cities. Renew Sust Energ Rev 55:273–287. https://doi.org/10.1016/j.rser.2015.10.133

Chávez A, Kennedy C, Chen B, Chertow M, Baynes T, Chen S, Bai X (2018) Understanding, implementing, and tracking urban metabolism is key to urban futures. In: Elmqvist T, Bai X, Frantzeskaki N et al (eds) Urban planet: knowledge towards sustainable cities. Cambridge University Press, Cambridge

Collier P, Jones P, Spijkerman D (2018) Cities are engines of growth: evidence from a new global samples of cities

Cugurullo F (2017) Exposing smart cities and eco-cities: Frankenstein urbanism and the sustainability challenges of the experimental city. Environ Plan A: Econ Space 50(1):73–92. https://doi.org/10.1177/2F0308518X17738535

Fan J-L, Kong L-S, Wang H, Zhang X (2019) A water-energy nexus review from the perspective of urban metabolism. Ecol Model 392:128–136. https://doi.org/10.1016/j.ecolmodel.2018.11.019

Gaigné C, Riou S, Thisse J-F (2012) Are compact cities environmentally friendly? J Urban Econ 72:123–136. https://doi.org/10.1016/j.jue.2012.04.001

Green J, Newman P (2022) Transactive electricity markets: transactive electricity: how decentralized renewable power can create security, resilience and decarbonization. In: Droege P (ed) Intelligent environments 2 – advanced systems for a healthy planet. Elsevier, Amsterdam

International Energy Agency (IEA) (2017a) World energy outlook-2017. OECD/IEA, Paris

International Energy Agency (IEA) (2017b) World energy outlook 2017: China. International Energy Agency, Paris

International Energy Agency and the United Nations Environment Programme (IEA, UNEP) (2018) 2018 global status report: towards a zero-emission, efficient and resilient buildings and construction sector. Global Alliance for Buildings and Construction (GlobalABC)

International Federation for Housing and Planning (2016) The role of technology in urban development. IFHP. https://www.ifhp.org/role-technology-urban-development. Accessed 17 Apr 2019

International Organization for Standardization (2015) Smart cities get on their bikes. ISO. https://www.iso.org/news/2015/09/Ref1998.html. Accessed 16 Apr 2018

IPCC (2018) Summary for policymakers. In: Masson-Delmotte V, Zhai P, Pörtner HO et al (eds) Global warming of 1.5°C. an IPCC special report on the impacts of global warming of 1.5°C above pre-industrial levels and related global greenhouse gas emission pathways, in the context of strengthening the global response to the threat of climate change, sustainable development, and efforts to eradicate poverty. World Meteorological Organization, Geneva.

IPCC (2022) Climate change 2022: impacts, adaptation and vulnerability. Reportno. Report Numberl, Date. Place Publishedl: Institutionl

IRENA (2017) IRENA cost & competitiveness indicators: rooftop solar PV. International Renewable Energy Agency, Abu Dhabi

IRENA (2018) Global energy transformation: a roadmap to 2050. International Renewable Energy Agency, Abu Dhabi

Jacobs J (1961) The death and life of great American cities. Random House, New York

Kalmykova Y, Rosado L, Patrício J (2016) Resource consumption drivers and pathways to reduction: economy, policy and lifestyle impact on material flows at the national and urban scale. J Clean Prod 132:70–80. https://doi.org/10.1016/j.jclepro.2015.02.027

Kaneda T, Greenbaum C, Kline K (2020) 2020 world population data sheet. Reportno. Report Numberl, Date. Place Publishedl: Institutionl

Kraus S, Richter C, Papigiannidis S, Durst S (2015) Innovating and exploiting enterpreneurial opportunities in smart cities: evidence from Germany. Creat Innov Manag 24(4):601–616

Kylili A, Paris AF (2015) European smart cities: the role of zero energy buildings. Sustain Cities Soc 15:86–95. https://doi.org/10.1016/j.scs.2014.12.003

Macháč J, Louda J, In DL (2016) Smart cities symposium Prague, Czech Republic, 26–27 May 2016. IEEE. https://doi.org/10.1109/SCSP.2016.7501030

Mahdavinejad MS, Rezvan M, Barekatain M, Adabi P, Barnaghi P, Sheth AP (2018) Machine learning for internet of things data analysis: a survey. Digit Commun Netw 4(3):161–175

Martinez-Balleste A, Perez-Martinez P, Solanas A (2013) The pursuit of Citizens' privacy: a privacy- aware smart city is possible. IEEE Commun Manag 51(6):136–141

Matan A, Newman P (2016) People cities: the Life and legacy of Jan Gehl. Island Press, Washington, DC

McKinsey & Company (2018) Smart city solutions: what drives citizen adoption around the globe? Mckinsey Center for Government

Meyer K, Newman P (2020) Planetary accounting. Springer, Berlin/Heidelberg

Mitra A, Mehta B (2011) Cities as the engine of growth: evidence from India. J Urban Plan Dev 137(2):171–183. https://doi.org/10.1061/(ASCE)UP.1943-5444.0000056

Motyka M, Slaughter A, Amon C (2018) Global renewable energy trends: solar and wind move from mainstream to preferred. Deloitte Development LLC, UK

Narayanan A, Mets K, Strobbe M, Develder C (2019) Feasibility of 100% renewable energy-based electricity production for cities with storage and flexibility. Renew Energy 134:698–709. https://doi.org/10.1016/j.renene.2018.11.049

Newman P (1999) Sustainability and cities: extending the metabolism model. Landsc Urban Plan 44(4):219–226. https://doi.org/10.1016/S0169-2046(99)00009-2

Newman P (2020a) Cool planning: how urban planning can mainstream responses to climate change. Cities 103:102651

Newman P (2020b) COVID, cities and climate: historical precedents and potential transitions for the new economy. Urban Sci 4(3):32

Newman P, Kenworthy J (1999a) Costs of automobile dependence: global survey of cities. Transp Res Rec 1670(1):17–26

Newman P, Kenworthy J (1999b) Sustainability and cities: overcoming automobile dependence. Island Press, Washington, DC

Newman P, Thomson G (2018) Urban fabrics and urban metabolism – from sustainable to regenerative cities. Resour Conserv Recycl 132:218–229

Newman P, Beatley T, Boyer H (2017) Resilient cities, Overcoming Fossil Fuel Dependence, 2nd edn. Island Press, Washington, DC

Oh D-S, Phillips F (2014) Technopolis: best practices for science and technology cities. Springer, London

Ong KCG (2017) Sustainable construction for Singapore's urban infrastructure – some research findings. Procedia Eng 171:14–21. https://doi.org/10.1016/j.proeng.2017.01.305

REN21 (2018) Renewables 2018 Global Status Report. REN21, Paris

Restrepo JDC, Morales-Pinzón T (2018) Urban metabolism and sustainability: precedents, genesis and research perspectives. Resour Conserv Recycl 131:216–224

Salingaros NA (2005) Towards a new urban philosophy. In: Hulsbergen E, Klaasen I, Kriens I (eds) Shifting sense – looing back to the future in spatial planning. Techne Press, Amsterdam, pp 265–280

Salingaros NA (2006) Compact city replaces sprawl. In: Graafland A, Kavanaugh L (eds) Crossover: architecture, urbanism, technology. 010 Publishers, Rotterdam, pp 100–115

Salingaros NA (2014) Complexity in architecture and design. Oz Journal 36:18–25

Samad MES, Hawken S, Sargolzaei S, Foroozanfa M (2019) Implementing citizen centric technology in developing smart cities: a model for predicting the acceptance of urban technologies. Technol Forecast Soc Chang 142:105–116. https://doi.org/10.1016/j.techfore.2018.09.012

Shahrokni H, Arman L, Lazarevic D, Nilsson A, Brandt N (2015) Implementing smart urban metaboism in the Stockholm royal seaport: smart city SRS. J Ind Ecol 19(5):917–929. https://doi.org/10.1111/jiec.12308

Shen L, Wu Y, Shuai C, Lu W, Chau KW, Chen X (2018) Analysis on the evolution of low carbon city from process characteristic perspective. J Clean Prod 187:348–360. https://doi.org/10.1016/j.jclepro.2018.03.190

Shepard W (2017) Why hundreds of completely new cities are being built around the world. Forbes https://www.forbes.com/sites/wadeshepard/2017/12/12/why-hundreds-of-completely-new-cities-are-being-built-around-the-world/#1380073e14bf. Accessed 17 Apr 2019

Slavova M, Okwechime E (2016) African smart cities strategies for Agenda 2063. Afr J Manag 2(2):210–229. https://doi.org/10.1080/23322373.2016.1175266

Steffen W, Richardson K, Rockström J et al (2015) Planetary boundaries: guiding human development on a changing planet. Science 347(6223):1259855

Stewart ID, Kennedy CA, Facchini A, Mele R (2017) The electric city as a solution to sustainable urban development. J Urban Technol 25(1):3–20

Stumpp E-M (2013) New in town? ON resilience and "resilient cities". Cities 32:146–166. https://doi.org/10.1016/j.cities.2013.01.003

Tosics I (2015) Growing cities: How to expand in a sustainable and integrated way? urbact: EU Regional Development Fund. https://urbact.eu/growing-cities-how-expand-sustainable-and-integrated-way. Accessed 17 Apr 2019

Tripathi V (2017) Achieving urban sustainability through Safe City. J Hum Ecol 59(1):1–9. https://doi.org/10.1080/09709274.2017.1356048

UN Habitats (2018) Brief on world cities day. https://unhabitat.org/wcd-2018

UN Population Division (2019) World urbanization prospects: the 2018 revision. Reportno. Report Numberl, Date. Place Publishedl: Institutionl

UNEP (2016) Global material flows and resource productivity. In: Schandl H, Fischer-Kowalski M, West J, Giljum S, Dittrich M, Eisenmenger N (eds) An assessment study of the UNEP international resource panel. United Nations Environment Program, Paris

United Nations (2016) The World's cities in 2016. Data Booklet, New York

United Nations (2018) 68% of the world population projected to live in urban areas by 2050, says UN. UN https://www.un.org/development/desa/en/news/population/2018-revision-of-world-urbanization-prospects.html. Accessed 11 Apr 2019

Van Winden W, van den Buuse D (2017) Smart city pilot projects: exploring the dimensions and conditions of scaling up. J Urban Technol 24(4):51–72. https://doi.org/10.1080/10630732.2017.1348884

World Bank (2010) Cities and climate change: an urgent agenda. Urban development series knowledge papers, vol 63704. Urban Development Series Knowledge Papers, Washington DC

World Commission on Environment and Development (1988) https://digitallibrary.un.org/record/139811?ln=en

Chapter 3
Smart Cities and Climate Change: How Smart City Can Help Cool the Planet

Check for updates

Abstract Smart cities are a critical part of the climate change agenda. The IPCC suggest that cities are a major part of the transition to zero carbon while enabling the transition to zero poverty as well. The key technologies for at least 80% of this change are now available with solar, battery and electric vehicles dramatically dropping in cost and exponentially increasing in adoption. However, they must be integrated into cities not just be part of individual households, and to do this will require integration of smart technology and associated smart systems. This chapter sets out how the integration can be done but recognises that the next major challenge is dealing with the socio-technical systems needed to creatively enable the mainstreaming of these technology packages.

3.1 Introduction

Technological innovation has accelerated in the last decade in terms of how the world will cope with the need to decarbonise the planet without destroying global economies. The future that is outlined below is emerging very quickly after the major downturn of COVID-19 collapse in the global economy (Newman 2020; Schwab and Malleret 2020; Roy 2020). This acceleration should not surprise us as scenarios of the future are not usually able to cope with the rapid change that we are now seeing but such changes have been characteristic of the last five waves of innovation following major economic collapses (Swart et al. 2004; Raskin 2016a). This chapter will show that smart technologies and systems are playing a key role on how the zero carbon city is emerging rapidly and how this can be enabled.

3.2 Waves of Innovation

Creative destruction 'is the essential fact about capitalism', wrote the Austrian economist Joseph Schumpeter in 1942 as the world was emerging from the Great Depression of the 1930s (Kondratieff 1984; Schumpeter 1930, 2015). New

Z. Allam, P. Newman, *Revising Smart Cities with Regenerative Design*, Cities and Nature, https://doi.org/10.1007/978-3-031-28028-3_3

technologies and processes continuously revolutionise the economic structure from within, 'incessantly destroying the old one, incessantly creating a new one'. This approach was formalised by a number of academics in the waves of innovation theory (Freeman and Soete 1997; Kondratieff 1984).

Change happens more quickly and creatively during times of economic disruption. Innovations that meet material and cultural needs accelerate, and structures preventing new, more efficient technologies weaken. As the old economy collapses, innovations 'cluster' to become the core of the new economy.

Over the past three centuries, there have been five great 'waves' of economic disruption and clustering. The first was driven by harnessing water power, the second by steam power, the third by coal and electricity, the fourth by oil and gas and the fifth by digital transformation. We are now at the start of the sixth great wave, driven by renewable energy combined with batteries, electromobility and Smart City technology. These waves are set out in Fig. 3.1 and Table 3.1.

The rapidly emerging innovations for a zero carbon agenda are outlined in Fig. 3.2 from the 2022 IPCC Mitigation Report showing their dramatic reductions in cost over the past decade and their exponential adoption into the economy.

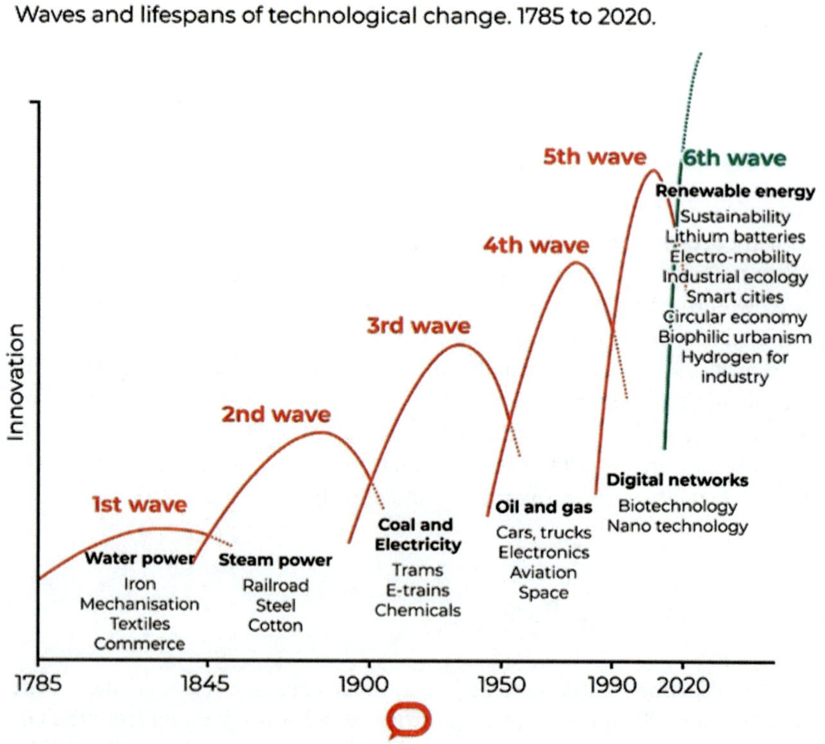

Fig. 3.1 The waves of innovation

Table 3.1 The waves of innovation following economic collapse in the past 250 years and what they meant for business models, energy and infrastructure and transport and urban form, with potential options for the next wave

Economic waves	Technological innovations emerging	Business model	Energy and infrastructure	Transport and city form
1. 1780s–1840s industrial revolution	Water power Iron Mechanisation Textiles Commerce	Small and cottage industries	Water power and horse power, canals and sailing ship ports; roads for carriages linking cities	Walking cities rapidly densifying from industry
2. 1840s 'hard times' and then Victorian prosperity	Steam power Railroad Steel Cotton	Cottage industries into large capital firms and factories	Wood and steam into train systems	Walking cities into rail-based linear urban development
3. 1890s great depression and then belle Epoque	Electricity Chemicals Internal combustion engine	Monopolistic Fordist Firms and factories	Coal and electric tram and train systems	Tram and train-based corridors
4. 1930s great crash and then Keynesian growth	Petrochemicals Aviation Electronics Space	Multinationals Modernism	Oil and freeways	Automobile-based urban sprawl
5. 1980s dot-com recession and then knowledge economy	Digital networks Biotechnology Information technology	'Smart cities' flexible specialisation and networked globalism	Superhighway and ICT systems	Revival of urban centres around knowledge economy
6. 2020s COVID-19 collapse and then green economy	Renewable energy Circular economy Smart city	'Regenerative urbanism' Global localisation based around precincts	Renewables with batteries, Electromobility, Smart City integration, hydrogen for industry, circular economy, biophilic urbanism	Re-localised centres Smart, distributed infrastructure Transit activated corridors fed by micromobility, EVs and active transport

Source: Newman (2020)

These rapid changes are however going to need to be integrated with smart technologies to enable them to mainstream. Without smart technologies, the solar-battery-EV revolution that is outlined in the above figure will stall and be rejected as they will not be able to create the economy-wide systems that are needed to transform how we make and use energy in buildings, industry and transport. This is the theme of the chapter.

The world needs to have all the best zero carbon technologies as outlined in the chapter, but it also needs to create employment that can enable the zero poverty

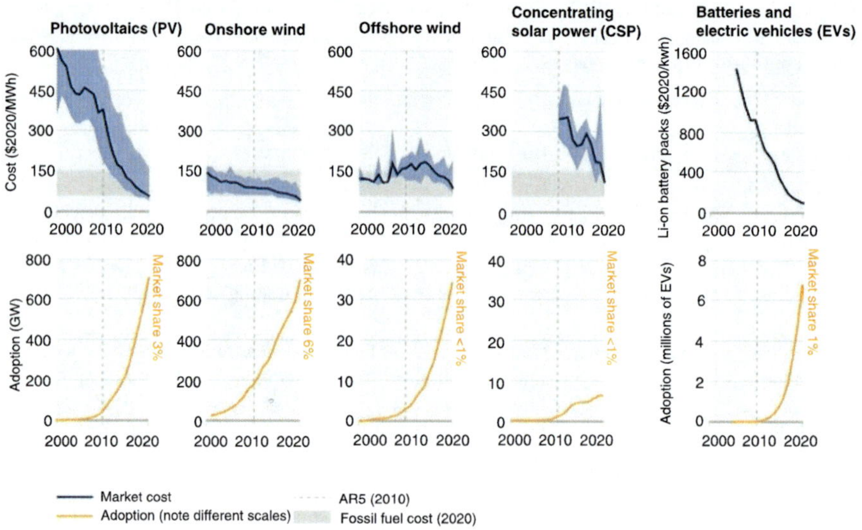

Fig. 3.2 Dramatic drops in costs of emerging technologies for zero carbon futures along with exponentially increasing adoption rates. (Source: IPCC 2022)

agenda. Although in the past these have been coupled so that economic growth caused increasing greenhouse emissions, the past decade has begun to show how these can be decoupled (Newman and Thomson 2018). Thus, the technologies outlined below are all cost-effective at enabling an agenda where zero carbon can be achieved along with zero poverty in new economic development strategies as part of the new economy (Newman et al. 2017). But they must be integrated into how cities and regions work, and hence it is important for both the zero carbon and zero poverty agendas that smart technologies are integrated into the solutions being adopted. This will require socio-technical change, not just technical change.

Cities will be the focus of such development as history has shown that is where economic growth can best be enabled particularly in the last 200 years (Glaeser 2011; Hall 2008). It is also where new technologies that are supportive of the zero carbon agenda are likely to be rapidly adopted, and that is where they are already beginning to be applied though there are also innovations occurring at the edge (rural and remote areas). Thus, the zero poverty and zero carbon agendas are likely to be rolled out in a synchronistic manner integrating the economic and the environmental agendas using smart technology systems. Some of the ways this can happen together will be outlined.

3.3 Zero Carbon Technologies

3.3.1 *Renewable Energy, Especially PV and Batteries*

The dramatic growth in renewable energy (solar and wind) has been due to these technologies quickly becoming the cheapest form of power that history has ever seen. They are also easy to mass produce and implement in most cities and most economic systems (Droege 2011; Scott 2020). Organisations that predict the future based on previous growth patterns have constantly got this wrong as these are disruptive innovations (Seba 2014; Newman et al. 2017) particularly rooftop solar as it enables very local production and consumption to be integrated (Newman 2017).

The new patterns of urbanism that are emerging around these systems are already showing why cities will become much more distributed into local areas of infrastructure management – but they will still fit into a citywide or region-wide grid system for equity and balance (Green and Newman 2017; Newton and Newman 2013). The rapid growth in solar has now moved into shared solar systems for medium- and high-density housing enabled by localised solar utilities (Newton et al. 2021).

However, this shared solar approach cannot be managed in an equitable and transparent way unless they also involve batteries and blockchain-based management. The WGV development in Perth has three shared solar systems among the 110 housing units that have been well studied (Byrne et al. 2021), and now the next stage of this area (called East Village) is also involving shared electric vehicle recharging opportunities, and the whole area is not just zero carbon but is 20% less costly (Byrne et al. 2021). Industrial estates with shared solar appear to be happening as well as rural and remote systems (Galloway and Newman 2014); the new Peel Industrial Estate in Perth has a shared microgrid solar system providing cheaper power options for businesses that locate there. All of these innovative developments are providing their solutions using integrated smart system technology (Green and Newman 2018) and are achieving zero carbon with all of the SDGs addressed (Jason et al. 2018).

The next agenda is how to achieve grid stabilisation, and this seems to be heading toward localised, community-scale batteries (Mey and Hicks 2019; Sproul 2019). These are becoming available for many other urban functions including electromobility which, as shown below, can be part of grid stabilisation. Gas turbines (and diesel backup in small grids) have been seen as necessary for grid stabilisation, but Li-ion batteries are now cost-effective at over 150 MW showing that they are now cheaper than gas turbines and more effective at providing a rapid peaking function (Denholm et al. 2021). Again these grids of the future will not work using solar and batteries and EVs unless they have smart systems.

Thus, 100% renewable power systems can now be built cost-effectively. The AEMO (2020) report on the Australian grid shows that it can be 75% renewable by 2025 pushed mostly by rooftop solar. Old coal and gas power in the COVID-19 downturn were being avoided as new solar and battery systems have marginal costs of zero; thus, the reductions in demand were carried by fossil fuel-based power, accelerating the decarbonisation transition. This has only hastened their phaseout. Such is a classic example of how exponential growth in renewables and batteries can be expected to take over power grids in the near post-COVID-19 future, especially driven by the large and growing sector of ethical investing and even the world's largest finance company Blackrock (Fink 2020) and other finance companies listed on the Climate 100+ website.

The investment in renewables is also enabling new manufacturing centres to emerge with much better local job creation if the products are created from localised renewables (Nahum 2020). Zero poverty associated with a just transition can also be a goal in areas based around coal, oil and gas, but the cost-effective manufacturing opportunities will rapidly shift to the best sites for renewables (Garnaut 2019). This is happening because the world's manufacturing is moving toward higher-quality raw materials that are better processed and utilised near where they are found rather than moving them to industrial areas with cheap labour.

3.3.2 Electromobility, Micromobility, Transit, Walkability and Active Transport

Electromobility is also a disruptive innovation that has emerged rapidly in recent years (IPCC 2022). With a growth rate of over 40% per year, predictions for how quickly the internal combustion engine (ICE) will be phased out are now coming down into the 2020s (IEA 2021a; Dia 2019). The reason electromobility is disruptive is similar to solar and wind: the vehicles are rapidly reducing in cost as battery packs have reduced from $1000/Kwh in 2010 to less than $200/Kwh in 2020 and because the technology is preferred for many other social and environmental reasons, especially cleaning up local air pollution (Dia 2019; Dia et al. 2019; Newman 2020). Bloomberg New Energy Finance (in McKinsey report 2017) now suggests capital costs will be equal to ICE in the mid-2020s and operational costs will be much less. The next electric vehicle types to roll off assembly lines in large numbers in the early 2020s were electric buses (Transport and Environmet 2018) and electric trucks (Limatainen et al. 2019) with some decades before ships and planes. Long distance ships and planes are likely to be fuelled by green hydrogen-based ammonia and synthetic jet fuel with some advanced biofuels (IPCC 2022).

Concern over the growth in consumption of battery minerals has been seen as a constraint on this rapidly growing Li-ion battery market (Olivetti et al. 2017). Responses to this show that these minerals are needed in very small quantities compared to iron and aluminium and the growth of assured supplies with ethical and transparent mining systems as well as growing recycling opportunities suggest these concerns are being addressed (IPCC 2022; Newman 2020; Newman et al. 2018; IEA 2021b; Lee et al. 2020; World Bank Group 2017). However, reduced consumption is a necessary agenda for multiple reasons and is addressed under Smart City-based demand management below.

Disruption is demand-driven, like the smartphone, but those cities adapting to new systems with solar-batteries-EVs are also seeing common-good outcomes that they can assist (Garnaut 2019). These public benefits are especially seen on how EVs are assisting electromobility in the new forms of micromobility and in new mid-tier electric transit which have multiple benefits in overcoming automobile dependence, the scourge of the last big economic wave after the 1930s (explored further below).

Micromobility Micromobility electric vehicles are all small, local transport technologies that support walking: electric tuktuks, electric bikes, electric motor bikes, electric scooters and electric skateboards that are becoming a major part of the EV revolution (Ajao 2019). The growth of these modes in Chinese cities (Gao et al. 2017) and places like Delhi (now with the fifth biggest recharge system in the world's cities; see StartUS, 2020) is driven by the need to reduce air pollution, but they are also part of the active transport movement that is showing substantial health benefits when reducing car dependence. Forty-six percent of car journeys in the USA go just 3 miles or less, which can simply be replaced by micromobility, and 30% of micromobility riders were doing just that during the COVID-19 shutdown period (Ajao 2019). The question is whether such options can be mainstreamed into the cities of the future. This is particularly interesting in view of the global energy crisis that has been prompted by the war in Ukraine.

Electric micromobility can also develop a range of new functions relevant to local centres (a growing focus for the new economy), which could include local delivery of online parcels. The rapid growth in parcel delivery vans has been a major part of growing traffic in the USA (Schaller 2018) and can be solved if parcels were delivered locally by micromobility small e-vans that could also be autonomous (as demonstrated in some cities during COVID-19 lockdown). They can be delivered regionally by trackless trams (see below) along corridors providing parcels to station precincts where local delivery vehicle hubs can be located.

All forms of electromobility need recharging, and in cities, these can become part of a new recharge hub or battery-storage precinct that is based strategically to support the grid balance needed to ensure universal access and resilience. These recharge hubs can be available to the multitude of micromobility vehicles that can be supportive of local economies and also provide the last-mile linkages for electric transit as it services a corridor of economic development.

Transit Perhaps the most significant innovation in electromobility in terms of common-good outcomes is the electrification of public transport (Glazebrook and Newman 2018). The electrification of heavy train and tram systems is a mature technology based on overhead catenaries, but new Li-ion batteries have revolutionised the electrification of buses into electric BRTs or trackless trams (Newman et al.

2019). Trackless trams are guided quickly and smoothly down main roads like a high-quality light rail by a smart system, powered by batteries on the roof. They are much cheaper than traditional tram systems as they do not need massive engineering to provide the steel tracks and overhead catenaries.

Such new e-transit systems will be critical to provide in the zero carbon city of the future as any city dominated by cars, whether petrol or battery-based, will have many problems. A strong e-transit system will help balance the congestion brought by EVs, alongside increased uptake on e-rideables and e-bikes. These trackless trams are now able to fit into cities by enabling the development of new precincts around stations due to their quiet, pollution-free accessibility that is able to replace the equivalent of six lanes of traffic. The urban regeneration can be used to help pay for the new transit, and the resulting station precincts can include recharge hubs for the battery-based transit and micromobility last-mile linkage. They are therefore enabling distributed infrastructure and supporting the development of a zero carbon and zero poverty city.

These transit innovations will enable creation of zero carbon corridors (Newman 2020) along main roads replacing the highly congested car dependence and enabling a series of urban regeneration sites to be developed around station precincts instead of sprawling cities at the urban fringe (Newman and Kenworthy 1989, 1999, 2015). This is likely to be a big part of future urban economic development strategies, detailed further below. Zero carbon corridors will also need significant help from smart systems to enable these shared, common-good outcomes in their operations.

Zero carbon corridors, also called twenty-first-century boulevards (Newman and Kenworthy 1999), can enable a zero carbon city to be built in stages through focusing on a new transit system along a main road (like a trackless tram; see Fig. 3.3) and then building zero carbon station precincts along the route as shown in Fig. 3.4. These zero carbon precincts can slowly absorb surrounding suburbs into their solar-based microgrids and enable the transition to spread across the city. They will be described more fully in Sect. 3.6.

Walkability and Active Transport The benefits of micromobility in enabling local centres to work without cars and to enable transit systems to work without the need for car-dependent corridors are certainly rapidly emerging over the past decades (Matan and Newman 2016; Gehl 2010). Transit was seriously damaged during COVID-19 but so was car traffic, so the growth of local walkability and active transport has been a global phenomenon with many cities building this into permanent change (Laker 2020; Davies 2020; Harris 2020). At the height of the pandemic in late April, across Scotland, the public transport was down by up to 95%, car use was down by 70% and cycling was up by 120% (Transport Scotland 2020). Rapid returns of traffic and transit have happened since then, but active transport has grown significantly (Earley and Newman 2021).

The co-benefits of active transport are very high, and if local economic development is facilitated, then it is also a part of the zero poverty agenda. Thus, closing streets to cars and building large numbers of cycle ways suitable for all

Fig. 3.3 The trackless tram system by CRRC and demonstrated in Zhuzhou, China. Source: Compliments of CRRC

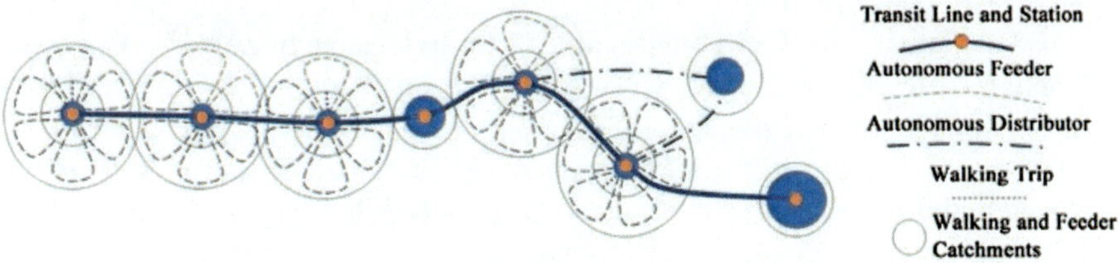

Fig. 3.4 A diagrammatic representation of accessibility of a twenty-first-century boulevard. (Source: Amended from Glazebrook and Newman 2018)

micromobility are likely to be a high priority for the recovery as announced in London (Quinn 2020) with a scenario of increasing active travel ten times. The Mayor and Transport for London said their goal was as follows:

- The 'rapid construction' of a strategic cycling network, using temporary materials, with new routes, aimed at reducing crowding on public transport.
- A 'complete transformation' of local town centres so that people can walk and cycle where possible, including widening footways on high streets so that people can safely queue outside shops.
- Reducing traffic on residential streets and creating 'low-traffic neighbourhoods'.

The project was given an immediate £2b on the basis that 'when the nation gets moving again it does so in a cleaner safer way' (Walawalkar 2020). Leading actors also began a special App to help with transport choices, began trials on e-scooters, a key part of micro-mobility, extra charging points for EV vehicles of all kinds, and accelerating new rail projects (Sky News 2020).

Re-localising the city like this becomes a strong positive outcome from the move to active transport with its support in micromobility and new electric transit systems as well as the localised power systems emerging from the solar-battery-based power systems to further the 'transformation' of local town centres. It is a sign that a new policy orientation has emerged from COVID-19. But it is also clear that all these innovations will not mainstream unless they are supported by smart systems.

3.3.3 Smart City-Based Demand Management

Smart city is an agenda which has rapidly grown in the twenty-first century but has many controversial aspects if it is used simply for surveillance and traffic congestion management largely failing due to the rebound effect (see Creutzig et al. 2016), which continues the problem of car dependence. In Box 3.1, we summarise the Smart City technologies that are all going to be needed for integrated solutions to the zero carbon and zero poverty agenda.

Box 3.1: Smart City Technologies and Their Relevance to Zero Carbon Cities

Information and Communication Technology (ICT). The use of ICT can help cities by providing real-time information on various dimensions, from solar to shared mobility; where the latter is beneficial for transit users and those using bikes and pedestrian pathways. ICT can help with building management and with information for transit users as well as ticketing and payment for transit or for road user charges as the world shifts away from fuel taxes due to the demise of ICE vehicles (Gossling 2017; Tafidis et al. 2017).

Internet of Things (IoT) Sensors. These sensors can be used for building management and road safety to ensure any vehicles do not lose their direction. Smart tyres equipped with sensors help to slightly improve fuel efficiency and reduce CO_2 emissions (Kavitha et al. 2018; Kubba and Jiang 2014). The IoT can be used to create safety for a fast-moving trackless tram, and its associated last-mile connectivity shuttles as part of a transit activated corridor (Newman et al. 2019).

Mobility as a Service (MaaS). New, app-based mobility platforms will allow for the integration of different transport modes (such as last-mile travel, shared transit, and even micro-transit such as scooters or bikes) into easy-to-use platforms. By integrating these modes, users will be able to easily navigate from A to B based on what modes are most efficient, and all necessary bookings or payments can be made through the one service. With Smart City planning, these platforms can steer more users toward shared and rapid transit

(contiuned)

Box 3.1 (continued)

(which should be the centrepiece of these systems), rather than encouraging more people to opt for the perceived convenience of booking a single-passenger ride. In low-density car-dependent cities, however, MaaS services such as the use of electric scooters/bikes are less effective as the distances are too long and they do not enable the easy sharing that can happen in dense station precincts (Jittrapirom et al. 2017).

Artificial Intelligence (AI) and Big Data Analytics. These technologies are used together to enable decisions about optimising use of solar and batteries in buildings and in what kind of transport planning is used down particular corridors. Options such as predictive congestion management of roads and freeways along with advanced shared transit scheduling can provide value to new and existing transit systems (Anda et al. 2017; Milne and Watling 2019; Toole et al. 2015).

Blockchain or Distributed Ledger Technology. Blockchain is presently being used as the basis of shared solar, battery and EV systems (Green and Newman 2018). Blockchain can also be the basis of MaaS or any local shared mobility as it facilitates shared activity. As the future city is going to have distributed solar energy, it can be applied to that and to how urban regeneration along a TAC can be sharing mobility opportunities, especially the payments for tickets on a transit system and its last-mile connectivity shuttles. This technology can also be used for road user charging along any corridor and by businesses accessing any services and in managing freight (Green and Newman 2017; Vujičić et al. 2020; Irannezhad 2020).

All of the innovations outlined above – solar, batteries, electromobility in vehicles of all shapes and sizes and other innovations such as circular economy, water and waste systems – have two key characteristics of relevance to this chapter: they work best as *localised systems*; and they work even better if *consumption is reduced.* Both can be helped significantly (or will not work at all unless implemented) by Smart City-based demand management. Thus, it is possible to imagine why Smart City technologies are an essential part of an integrated package for future cities of solar-battery-electromobility-smart systems. And as outlined in the last chapter, they must be integrated and delivered together with zero carbon technologies.

Localised Systems. The new Smart City technologies include an ability to enable any system to learn and optimise itself through artificial intelligence (AI) or machine learning. Many functions of AI have been envisioned to help the zero carbon agenda (Rolnick et al. 2022), but there is one that is just emerging which can assist urban regeneration of precincts. As new suburbs, new industrial estates, new office blocks and new villages are built or refurbished, they can be enabled through sensors to manage their energy, water, waste and transport, to be continuously learning from its users. Thus, the centres become something like a set of neural networks that are constantly

(contiuned)

Box 3.1 (continued)

improving the ecosystem in which they operate. The roles of recharge hubs and delivery vehicle hubs can all be optimised along with many other new localised services.

The local precincts set up with such simple technology will be highly efficient and can be optimised to share equitably and enable job creation through enterprise facilitation in local communities (Sirolli 1995, 2012). The goal is to provide a cost-effective liveability and flourishing economy while reducing consumption of resources. Welfare for people with disabilities or those in aged care can also be improved with these kinds of infrastructure systems. Localised smart systems can be managed to provide different solutions for the different kinds of places across a city and its region as outlined in Sect. 3.5 on socio-technical transitions.

Reduced Consumption. Resource conservation is a critical part of the zero carbon and zero poverty agenda (Creutzig et al. 2016, 2018; IPCC 2022). This is necessary for both better environmental, employment and economic outcomes and making the transition quicker and easy. Smart technologies can be used to reduce consumption through assisting behaviour change and demand management systems (Creutzig et al. 2018). There is evidence that behaviour can be influenced by direct social interventions by government programs (Abrahamse and Steg 2013; O'Brien et al. 2019), but mostly they fail unless they are combinations of social and technological change which provides new knowledge-based systems (Eon et al. 2019), or they enable infrastructure to work better through enhancing lifestyles in low-carbon urban form or changes like re-localisation (Newman and Kenworthy 2015). Smart technologies combined with demand management programs are likely to be much more effective as an integrated package (IPCC 2022).

These innovations are called Smart City-based demand management (Pears and Moore 2019). They enable householders and businesses to understand what they are consuming at any point in time with mobile phone apps and displays in homes and offices, with simple programmable options that build in the optimal efficiencies for use of energy, water, other materials and transport (Byrne et al. 2019). These can be simple apps such as the ClimateClever Home calculator (https://www.climateclever.org/homes) and can also be built into new houses and precincts from the start as part of a zero carbon home or precinct. Zero carbon transport can use these new kinds of smart system technologies as part of new precincts that are built around new transit systems with walkable environments (Allam and Newman 2018).

Industry can be part of this agenda through industrial estates that become learning environments as they share resources and wastes through industrial ecology (Hangroves and Smith 2005; Galloway and Newman 2014). But there are big industries that are not yet part of the zero carbon agenda as so far cement and steel production and most mineral processing need to use coal and gas. However, hydrogen-based options combined with solar-electric systems are emerging and could rapidly move ahead in the new economy (IPCC 2022).

3.4 Socio-technical Transitions

The long-wave theory outlined in Fig. 3.1 and Table 3.1 shows how innovations develop. These transitions have been outlined by economists, but a different kind of literature has tried to show that technological change is not just a change of economic models but must include social, cultural and political change, or else barriers will prevent their mainstreaming. These two transition theories are the transition management work by Frantzeskaki et al. (2018, 2019), and Malekpour et al. (2020), as well as the socio-technical transition theory of Geels (2005, 2002, 2011) and many followers. Both groups are seeking to find how technological innovations can be enabled more effectively as part of cities, with all their cultural, social and institutional barriers.

Socio-technical transition research has been very helpful for decision-makers and professionals to see how they need to bring whole-system structures into their plans for delivering innovations. They are mostly, however, dealing with a world that is pre-COVID-19, when there was significant momentum behind the economic system at that time, with lock-in based on investments in infrastructure, buildings, people and manuals of how things should be done (Mattauch et al. 2015). Economic collapses mean there is a discontinuity which is far more significant for opening up economies, cities and culture to rapid change than the incremental understandings in most scenario work (Raskin 2016b). Assets become stranded, and investment shifts to the new options, with potential to last longer. Hence, these barriers may be significantly reduced as the world starts up again because economic collapse may have 'cleaned the slate' as Le Corbusier said as he pushed the modernist agenda after the 1930s collapse (Flint 2014). However, the barriers may also remain and make change very difficult.

3.5 Application of Socio-technical Transitions to Zero Carbon Corridors

In this section, an example is provided of how socio-technical transition theory can be applied to integrate solar, battery and electric vehicles with smart systems in order to create a zero carbon city. As outlined in Sect. 3.4, the potential of new zero carbon transit technology in trackless trams can be used to help create zero carbon corridors. Zero carbon precincts around stations are necessary for zero carbon corridors and will be very attractive to developers because they are value enhancing and are risk-proof investments with a new iron-clad ability to raise finance which requires zero carbon to be the fundamental outcome of the project (see Climate Action 100+). New precinct-scale systems are the best way to introduce the modular carbon-reducing technologies like solar, batteries, new small-scale water and waste systems and the new local electric transport systems also outlined above (Thomson et al. 2019; Newton and Taylor 2019).

One of the most important integrative features of these new urban systems is that each precinct/station can become a recharge hub for the trackless tram and the e-micromobility which will provide a clear destination point along the net zero corridor with people feeding into and out of the station from around the precinct (Newman, 2018) (see Fig. 3.4).

In practice, such a net zero corridor will allow electric private cars to move down the central lanes (with the trackless tram operating in a dedicated lane), and when nearing stations, private vehicles will be required to slow down to enable the integration of the walkable environment adjacent to the station precinct. A zero carbon corridor can have a chain of precincts along the corridor that all use these new technological advancements for regenerating a city's infrastructure and buildings.

Given the interconnected nature of a trackless tram style zero carbon corridor, different station precincts along a corridor can have their own distinct amenity and purpose, such as a focus on healthcare, education, dining, entertainment or offices. This special quality of place-based urbanism is fundamental to any urban regeneration as it will be the need for the integration of nature (both as part of urban design and for aesthetics), effective waste management (such as waste minimisation and enhanced collection of recyclable items) and green buildings (Caldera et al. 2019). This will add value to the development.

But most importantly, the necessary uplift in value that can release the funding/financing of a series of urban regeneration precincts will only happen if there is a strong and competitive transit system feeding the residents, workers and visitors to the precinct. Each precinct will therefore be an opportunity to show how they can use new technology in their project and most importantly how they can link into innovative transit systems.

It is clear that all forms of vehicles will soon be electric, both private and shared vehicles, including those used for last-mile linkages. Thus, this presents an opportunity for each of the precincts to provide on-site energy generation (using solar technology), storage (using battery technology) and charging infrastructure (both high speed and low speed). Thus, this shifts the focus of the net zero corridor from simply a transport route to being a key part of the surrounding electricity grid, presenting both challenges and opportunities.

The understanding that very soon vehicles will be a key part of the electricity grid has given rise to the concept of 'vehicle-to-grid' or V2G, which describes a system where electric vehicles are connected to the grid for mutual benefit (Uddin et al. 2018). Vehicles have large batteries so around 100,000 vehicles can provide 500 MW of storage but large buses and trucks will provide much more, and if they are in a fleet, they can be attached to the grid through their depots and provide extra services to the grid when needed. Once connected, a range of services can be designed to allow electric vehicles to contribute to storage and grid stabilisation, increasing grid efficiency, stability and reliability (Yilmaz and Krein 2012). The physical connectivity between vehicles and the electricity grid will likely allow for discharging of power from transit and other vehicles when not in use, to respond to times of peak demand, referred to as 'peak shaving' (Ehsani et al. 2012). These provide six potential services that electric vehicles will likely provide grids in the future ranging from peak shaving for between 15 min and 2 h down to assisting the starting of electric motors that require high-intensity electricity for a short period of approximately 15 s.

What this adds to grid support is the potential innovation to create 'recharge hubs' at station precincts which support urban regeneration as well as grid stability. These will be able to help all kinds of electric vehicles recharge, especially micromobility feeding into the electric transit that is suitable for constrained spaces. The new precinct developments can be built with solar-PV on roof and wall space and extend out as far as is needed to ensure sufficient power can be provided locally.

All forms of electromobility need recharging, and in cities, these can become part of recharge hubs at station precincts that are managed strategically to support the grid to ensure universal access and resilience. These recharge hubs can be available to the multitude of micromobility vehicles that can be supportive of local economies and also provide the last-mile linkages for electric transit as it services a corridor of economic development.

Moving from theory to practical application in this area is still a focus of research and demonstration efforts, but it is very obvious that it is not just technical, but it is socio-technical. From the above outline of the concept of an integrated approach using smart systems to integrate solar, battery and electric vehicles, the following questions need to be answered to enable such zero carbon corridors to be mainstreamed:

1. What kind of design and financial criteria will be needed to replace normal procurement when approaching finance companies to put together a zero carbon corridor?
2. Are utilities wanting help with their grids along the proposed zero carbon corridors, and will the proposed concept fit these needs?
3. How will the traffic management be worked out along the corridors to enable the trackless tram to have right-of-way?
4. How can a zero carbon corridor integrate amenity and services and place qualities into their fabric so that local communities and local governments will want to help enable it?
5. What kind of power recharge systems can be integrated into the precinct to enable all kinds of electric vehicles to be serviced, what space is needed and how should these services be managed?
6. Are V2G barriers being removed in the zero carbon corridors, and are the benefits being passed on to the precinct?
7. How are the microgrids that are linking precinct scale solar, battery and electric vehicle recharge going to be managed, and what smart technologies can help enable this?
8. Are the professional manuals for best practice urban development changing to incorporate these new opportunities?

The questions above are likely to be very challenging for any level of government and for all the professionals involved. They are typical of the kind of socio-technical transitions issues that will enable the mainstreaming of smart, sustainable cities. Entrepreneurial developers and professionals will be needed to provide such answers and enable demonstrations to help walk the old systems of governance into the new era. They will all need Smart City technology fully integrated into the zero carbon packages.

3.6 Conclusions

The benefits to the economy of integrating renewable energy generation, batteries, electromobility and Smart City technologies at local levels are emerging rapidly as they are now highly commercial and in demand. They will lead to a cleaner, greener economy with many more new jobs and will give us hope as we seek to achieve a planet that is liveable into the future. The cities that take up these smart, sustainable options will be the leading forces that will drive the transition to zero carbon and zero poverty economies. They will be highly competitive places to live and work.

Based on IPCC (2022) estimates, the new technologies have the potential to reduce the use of fossil fuel by 80% in a decade. Eliminating the last 20% will be harder. This needs new processes not yet commercial. The gas and coal used in industrial processes such as steel production and mineral processing, and fossil fuels used for long-haul road, sea and air transport, need to be converted into a mixture of solar-electric and green hydrogen-based systems, most of which are still at demonstration phase. This will likely take a decade or more to enable commercial technology options as well as socio-technical processes to be worked through. All these changes will depend on smart systems as they emerge into commercial uptake and have to be integrated into grids in the same way solar-batteries-EV are now doing.

The only thing that will put the brake on these technologies becoming the core of the new economy sooner rather than later are backward-looking government policies that seek to prop up an obsolete fossil fuel economy by not changing the socio-technical systems that are preventing so many good opportunities as set out in this chapter. There will also always be a need for creative, entrepreneurial businesses and professionals that can demonstrate the benefits of integrating smart technologies into climate change technologies and who are prepared to creatively solve the socio-technical problems of implementing new zero carbon-zero poverty technologies in an integrated way using smart systems.

References

Abrahamse W, Steg L (2013) Social influence approaches to encourage resource conservation: a meta-analysis. Glob Environ Chang 23:1773–1785

Ajao A (2019) Mobility: here's everything you need to know. Available at: https://www.forbes.com/sites/adeyemiajao/2019/02/01/everything-you-want-to-know-about-scooters-and-micro-mobility/?sh=55f19b7f5de6

Allam Z, Newman P (2018) Redefining the smart city: culture, metabolism and governance. Smart Cities 1:4

Anda C, Erath A, Fourie PJ (2017) Transport modelling in the age of big data. Int J Urban Sci 21:19–42

Byrne J, Berry S, Eon C (2019) Transitioning to net zero homes—learnings from the CRC's high-performance housing living laboratories. In: Newton P, Prasad D, Sproul A et al (eds) Decarbonising the built environment: charting the transition. Palgrave Macmillan, Singapore

Byrne B, Liu J, Lee M et al (2021) The carbon cycle of Southeast Australia during 2019–2020: drought, fires, and subsequent recovery. AGU Adv 2:e2021AV000469

Caldera HTS, Desha C, Dawes L (2019) Evaluating the enablers and barriers for successful implementation of sustainable business practice in 'lean' SMEs. J Clean Prod 218:575–590

Creutzig F, Fernandez B, Haberl H et al (2016) Beyond technology: demand-side solutions for climate change mitigation. Annu Rev Environ Resour 41:173–198

Creutzig F, Roy J, Lamb WF et al (2018) Towards demand-side solutions for mitigating climate change. Nat Clim Chang 8:260–263

Davies A (2020) The pandemic could be an opportunity to remake cities. Available at: https://www.wired.com/story/pandemic-opportunity-remake-cities/

Denholm P, Arent D, Baldwin S et al (2021) The challenges of achieving a 100% renewable electricity system in the United States. Joule 5

Dia H (2019) Rethinking urban mobility: unlocking the benefits of vehicle electrification. In: Newton P, Prasad D, Sproul A et al (eds) Decarbonising the built environment: charting the transition. Palgrave Macmillan, Singapore

Dia H, Taylor M, Stone J et al (2019) Low carbon urban mobility. In: Newton P, Prasad D, Sproul A et al (eds) Decarbonising the built environment: charting the transition. Palgrave Macmillan, Singapore

Droege P (2011) Urban energy transition: from fossil fuels to renewable power. Elsevier Science, Amsterdam

Earley R, Newman P (2021) Transport in the aftermath of COVID-19: lessons learned and future directions. Sci Res 11:109–127

Ehsani M, Falahi M, Lotfifard S (2012) Vehicle to grid services: potential and applications. Energies 5:4076–4090

Eon C, Breadsell J, Morrison G et al (2019) Shifting home energy consumption through a holistic understanding of the home system of practice. In: Newton P, Prasad D, Sproul A et al (eds) Decarbonising the built environment: charting the transition. Palgrave Macmillan, Singapore

Fink S (2020) Ethical dilemmas in COVID-19-19 medical care: is a problematic triage protocol better or worse than no protocol at all? Am J Bioeth 20:1–5

Flint A (2014) Modern man: the life of Le Corbusier, architect of tomorrow. Amazon Publishing, New York

Frantzeskaki N, Hölscher K, Bach M et al (2018) Co-creating sustainable urban futures—a primer on transition management in cities. Springer, Berlin

Frantzeskaki N, Hölscher K, Holman IP et al (2019) Transition pathways to sustainability in greater than 2 °C climate futures of Europe. Reg Environ Chang 19:777–789

Freeman C, Soete L (1997) The economics of industrial innovation. MIT Press, Boston

Galloway D, Newman P (2014) How to design a sustainable heavy industrial estate. Renew Energy 67:46–52

Gao Y, Kenworthy J, Newman P et al (2017) Transport and mobility trends in Biejing and Shanghai: implications for urban passenger transport energy transitions worldwide. In: Droege P (ed) Urban energy transitions, 2nd edn. Elsevier, Berlin

Garnaut R (2019) Superpower: Australia's low-carbon opportunity. Schwartz Books Pty. Limited, Collingwood

Geels FW (2002) Technological transitions as evolutionary reconfiguration processes: a multi-level perspective and a case-study. Res Policy 31:1257–1274

Geels FW (2005) Technological transitions and system innovations: a co-evolutionary and socio-technical analysis. Edward Elgar, Cheltenham

Geels FW (2011) The multi-level perspective on sustainability transitions: responses to seven criticisms. Environ Innov Soc Trans 1:24–40

Gehl J (2010) Cities for people. Island Press, Washington, DC

Glaeser E (2011) Triumph of the city: how our greatest invention makes us richer, smarter, greener, healthier, and happier. Penguin Publishing Group, London

Glazebrook G, Newman P (2018) The City of the future. Urban Plan 3:1–20

Gossling S (2017) The psychology of the car: automobile admiration, attachment, and addiction. Elsevier Science

Green J, Newman P (2017) Citizen utilities: the emerging power paradigm. Energy Policy 105:283–293

Green J, Newman P (2018) Planning and governance for decentralised energy assets in medium-density housing: the WGV Gen Y case study. Urban Policy Res 36:201–214

Hall P (2008) Cities and civilization. Pantheon Books, New York

Hangroves K, Smith M (2005) The natural advantage of nations: business opportunities, innovation and governance in the 21st century. The Natural Edge Project: Earthscan, London

Harris S (2020) Post lockdown London: parking spaces and traffic lanes to vanish to encourage walking and cycling. Available at: https://www.itv.com/news/london/2020-05-06/post-lockdown-londonparking-spaces-and-traffic-lanes-to-vanish-to-encourage-walking-and-cycling/

IEA (2021a) Global EV outlook 2021. Technology report. IEA, Paris

IEA (2021b) The role of critical minerals in clean energy transitions: part of world energy outlook. IEA, Paris

IPCC (2022) Climate change 2022: mitigation for climate change. In: Working Group III contribution to the Sixth Assessment Report of the Intergovernmental Panel on Climate Change. IPCC

Irannezhad E (2020) Is blockchain a solution for logistics and freight transportation problems? Transport Res Proc 48:290–306

Jason W, Babaeff T, Breadsell J et al (2018) WGV: an Australian urban precinct case study to demonstrate the 1.5 °C agenda including multiple SDGs. Urban Plan 3:64–81

Jittrapirom P, Caiati V, Feneri AM et al (2017) Mobility as a service: a critical review of definitions, assessments of schemes, and key challenges. Urban Plan 2:13

Kavitha M, Raja M, Perumal P (2018) Evaluation of probiotic potential of Bacillus spp. isolated from the digestive tract of freshwater fish Labeo calbasu (Hamilton, 1822). Aquacult Rep 11:59–69

Kondratieff N (1984) The major economic cycles, translated in 1984 as the long wave cycles. E P Dutton, New York

Kubba AE, Jiang K (2014) A comprehensive study on technologies of tyre monitoring systems and possible energy solutions. Sensors (Basel) 14:10306–10345

Laker L (2020) World cities turn their streets over to walkers and cyclists. Available at: https://www.theguardian.com/world/2020/apr/11/world-cities-turn-their-streets-overto-walkers-and-cyclists?CMP=Share_AndroidApp_Tweet#maincontent

Lee J, Bazilian M, Sovacool B et al (2020) Reviewing the material and metal security of low-carbon energy transitions. Renew Sust Energ Rev 124:109789

Liimatainen H, van Vliet O, Aplyn D (2019) The potential of electric trucks – an international commodity-level analysis. Appl Energy 236:804–814

Malekpour S, Walker WE, de Haan FJ et al (2020) Bridging decision making under deep uncertainty (DMDU) and transition management (TM) to improve strategic planning for sustainable development. Environ Sci Pol 107:158–167

Matan A, Newman P (2016) People cities: the life and legacy of Jan Gehl. Island Press, Washington, DC

Mattauch L, Creutzig F, Edenhofer O (2015) Avoiding carbon lock-in: policy options for advancing structural change. Econ Model 50:49–63

Mey F, Hicks J (2019) Community owned renewable energy: enabling the transition towards renewable energy? 65–82

Milne D, Watling D (2019) Big data and understanding change in the context of planning transport systems. J Transp Geogr 76:235–244

Nahum D (2020) Powering onwards: Australia's opportunity to reinvigorate manufacturing through renewable energy. The Australia Institute Centre for Future Work, Canberra

Newman P (2017) The rise and rise of renewable cities. Renew Energy Environ Sustain 2:2017

Newman P (2018) The theology of sustainability practice. In: Marques J (ed) Handbook of engaged sustainability. Springer, Cham

Newman P (2020) COVID-19, CITIES and CLIMATE: historical precedents and potential transitions for the new economy. Urban Sci 4:32

Newman PWG, Kenworthy JR (1989) Gasoline consumption and cities. J Am Plan Assoc 55:24–37

Newman P, Kenworthy J (1999) Costs of automobile dependence: global survey of cities. Transp Res Rec 1670:17–26

Newman P, Kenworthy J (2015) Automobility and resilience: a global perspective. In: The urban political economoy and ecology of automobility driving cities, driving inequality, driving politics walks. Routledge, pp 28–58

Newman P, Thomson G (2018) Urban fabrics and urban metabolism – from sustainable to regenerative cities. Resour Conserv Recycl 132:218–229

Newman P, Beatley T, Boyer H (2017) Resilient cities, second edition: overcoming fossil fuel dependence. Island Press, Washington, DC

Newman P, Wills R, Edwards C et al (2018) Lithium Valley: building the case to establish Western Australia as a manufacturer of energy metals and power storage. Regional Development Australia, Kwinana Industries Council and Curtin University, Perth

Newman P, Hargroves K, Davies-Slate S et al (2019) The trackless tram: is it the transit and city shaping catalyst we have been waiting for? J Transport Technol 9:31–55

Newton P, Newman P (2013) The geography of solar PV and a new low carbon urban transition theory. Sustainability 5:2537–2556

Newton P, Taylor M (2019) Precinct design assessment: a guide to smart sustainable low carbon development. Co-operative Research Centre for Low Carbon Living

Newton P, Newman P, Glackin S et al (2021) Greening the greyfields: new models for regenerating the middle suburbs of low-density cities. Palgrave Macmillan, Melbourne

O'Brien LV, McNeil IM, Bruce G et al (2019) Social psychological guide for tranformation into low carbon living. In: Newton P, Prasad D, Sproul A et al (eds) Decarbonising the built environment: charting the transition. Palgrave Macmillan, Singapore

Olivetti E, Ceder G, Gaustad G et al (2017) Lithium-ion battery supply chain considerations: analysis of potential bottlenecks in critical metals. Joule 1:229–243

Pears A, Moore T (2019) Decarbonising household energy use: the smart meter revolution and beyond. In: Newton P, Prasad D, Sproul A et al (eds) Decarbonising the built environment: charting the transition. Palgrave Macmillan, Singapore

Quinn B (2020) London cycling could increase tenfold after lockdown. Available at: https://www.newsbreak.com/news/0OyjFBww/london-cycling-could-increase-tenfold-after-lockdown-says-tf

Raskin P (2016a) The great transition to planetary civilization. Tellus Institute, Boston

Raskin P (2016b) Journey to earthland: the great transition to planetary civilization. Tellus Institute, Boston

Rolnick D, Donti PL, Kaack LH et al (2022) Tackling climate change with machine learning. Comput Soc 55:1–96

Roy A (2020) The pandemic is a portal. Available at: https://www.ft.com/content/10d8f5e8-74eb-11ea-95fe-fcd274e920ca

Schaller B (2018) The new automobility: Luff, Uber and the future of American cities. Schaller Consulting, New York

Schumpeter JA (1930) Business cycles. Random House, New York

Schumpeter JA (2015) Capitalism, socialism and democracy. Routledge Classics, New York

Schwab K, Malleret T (2020) The great reset. World Economic Forum Publishing, Geneva

Scott M (2020) Solar and wind costs continue to fall as power becomes cleaner. Available at: https://www.forbes.com/sites/mikescott/2020/04/30/solar-and-wind-costs-continue-to-fall-as-power-becomes-cleaner/?sh=4ffe655e785f

Seba T (2014) Clean disruption of energy and transportation. Clean Planet Ventures, Silicon Valley

Sirolli E (1995) Ripples from the Zambezi: passion, entrepreneurship, and the rebirth of local economies. New Society Publishers, Vancouver

Sirolli E (2012) How to start a business & ignite your life: a simple guide to combining business wisdom with passion. Square One Publishers, New York

Sky News (2020) UK government unveils £2bn cycle and walk to work plan for COVID-19-19. *COVID-19-19*. YouTube. https://www.youtube.com/watch?v=UlteeyvX-3g

Sproul A (2019) Rooftop photovoltaics: distributed renewable energy and storage (or low-cost PV changes everything). In: Newton P, Prasad D, Sproul A et al (eds) Decarbonising the built environment: charting the transition. Springer, Singapore, pp 53–64

Swart RJ, Raskin P, Robinson J (2004) The problem of the future: sustainability science and scenario analysis. Glob Environ Chang 14:137–146

Tafidis P, Macedo E, Coelho MC et al (2017) Exploring the impact of ICT on urban mobility in heterogenic regions. Transport Res Proc 27:309–316

Thomson G, Newton P, Newman P (2019) Sustainable precincts: transforming Australian cities one neighbourhood at a time. In: Newton P, Prasad D, Sproul A et al (eds) Decarbonising the built environment: charting the transition. Springer, Singapore, pp 211–225

Toole JL, Colak S, Sturt B et al (2015) The path most traveled: travel demand estimation using big data resources. Transport Res C Emerg Technol 58:162–177

Transport and Environmet (2018) Annual report 2018. In: Todts W (ed) Transport and environment

Transport Scotland (2020) COVID-19-19 transport and trend data- 20–26 April 2020. Transport Scotland, Edinburgh

Uddin K, Dubarry M, Glick MB (2018) The viability of vehicle-to-grid operations from a battery technology and policy perspective. Energy Policy 113:342–347

Vujičić S, Hasanspahić N, Car M et al (2020) Distributed ledger technology as a tool for environmental sustainability in the shipping industry. J Mar Sci Eng 8:366

Walawalkar W (2020) UK plans £250m boost for cycle lanes and fast-track e-scooter trials. Available at: https://amp-theguardian-com.cdn.ampproject.org/c/s/amp.theguardian.com/uk-news/2020/may/09/uk-to-invest-cycle-lanes-coronavirus-air-pollution

World Bank Group (2017) The growing role of minerals and metals for a low carbon future. World Bank Publications, Washington, DC

Yilmaz M, Krein PT (2012) Review of benefits and challenges of vehicle-to-grid technology. In: 2012 IEEE energy conversion congress and exposition (ECCE). IEEE, pp 3082–3089

Chapter 4
Smart Cities and Regenerative Approaches: How Design with Smart City Tech Can Regenerate Cities

Abstract Regenerative approaches are seen as the third paradigm in managing economic development to be more socially and environmentally acceptable. Regenerative design is thus about repair of previous damage as part of the process of urban development. Such regenerative design is applied to each of five urban fabric types using the new technologies and design systems that are driving the post-COVID-19 economy. These are then shown to be absolutely requiring Smart City systems to make them mainstreamed into transforming urban economies.

4.1 Introduction

The future is rapidly emerging as set out in Chap. 3, and it clearly will depend on new technologies, especially solar, batteries and electric vehicles, as the climate change agenda is driving this and the vast majority of finance is dependent on achieving zero carbon outcomes with zero poverty as the simplified goal for the rest of the SDGs. But as we showed there, the mainstreaming of these emerging technologies depends heavily on their integration using smart technology and smart systems. Thus, we need to develop new ways of assisting Smart Cities to grow and create new options for each part of the city and each part of the global world. They must go beyond the simple branding exercises so common in the past 20 years to addressing real issues prompted by realities such as urbanisation and fast-growing urban population, the COVID-19 pandemic and the crisis resulting from geopolitical conflicts.

This chapter will take the zero carbon-zero poverty agenda a step further. It will show that not only do we need to regenerate cities as in the ancient tradition of giving new life to old urban structures and fabrics (Hall 2008) but that we need to be guided by a new kind of goal that is emerging called regenerative development. Regenerative design has been created by architects in the past 20 years as they sought to create urban fabric that repaired the natural environment around them. The best examples of this approach have shown that natural systems of soils, water and biodiversity habitat have been regenerated through design that enables ecological and landscape features to be prioritised with a focus on repair and rebuilding

Z. Allam, P. Newman, *Revising Smart Cities with Regenerative Design*, Cities and Nature, https://doi.org/10.1007/978-3-031-28028-3_4

their systems (Hes and Du Plessis 2014). This has shifted in more recent times to see how regenerative development can be applied to cities in a range of other ways (Thomson and Newman 2018). This chapter will describe how regenerative development is emerging and how smart technology and smart systems are critical to this transition.

4.2 Regenerative Approaches

4.2.1 New Business Models for the New Economy

Table 4.1 showed that each wave of innovation leading to a new economic era was characterised by a new business model. So, what business model could enable the cities and regions of the world to flourish through their investments and management systems? Two key ideas stand out: globalised localisation and regenerative ecosystems.

Globalised Localisation The global economy has become a key part of the twentieth century, and it is unlikely to lose its appeal as human curiosity for how the rest of the world works will always be there. It is also a world committed to zero poverty by 2030 through the SDG1, which have been adopted by every nation. This goal has been damaged by the impacts of COVID-19 like the near collapse of global economy and the disruptions of flow of money and the supply chain for major commodi-

Table 4.1 Technological approaches for the new economy and how they relate differently to five types of urban fabric

Approaches outcomes	Walking fabric	Transit fabric	Automobile fabric	Peri-urban and rural village fabric	Remote settlement fabric
Renewable energy (PV-B)	✓	✓✓	✓✓✓	✓✓✓	✓✓✓✓
Electromobility	✓ Micromobility	✓✓✓ Transit and micromobility	✓✓✓ Cars	✓✓ Cars and farm vehicles	✓ Off-road vehicles
Walkability and active transport	✓✓✓	✓✓	✓	✓	
Smart city demand mgt	✓✓✓	✓✓	✓✓	✓✓	✓✓
Hydrogen for industry			✓	✓✓✓	✓✓✓✓
Circular economy	✓	✓✓	✓✓✓	✓✓✓✓	✓✓
Biophilic urbanism	✓✓✓	✓✓	✓		
Permaculture		✓	✓✓	✓✓✓	✓

ties especially food due to the Ukraine war prompted by Russia's invasion. However, the notion of a global zero poverty economy will still drive the global agenda and requires each local economy to be picked up and given a new boost of localised development. This is appearing now to be associated with the zero carbon agenda through global finance and the UN's climate change agenda.

Zero carbon needs to be achieved by 2050, and the IPCC (2022) suggests this needs to be 50% in place by 2030. As shown in Chap. 3, the zero carbon agenda is fundamental to the new economy and is shaping the nature of global markets – they are becoming more localised. The reason for this is that the technology is much localised and is often called 'distributed' rather than 'centralised'. Solar and wind are very different across the globe and very variable. However, they are the cheapest form of energy now, so this will drive them into being developed in different ways in different places. They can also be applied in different parts of cities in different ways as will be set out in this chapter.

The application of renewable energy along with batteries and other storage was shown in Chap. 3 to be highly dependent on smart technology and smart systems that integrated these resources into achieving local outcomes as part of a wider system. This is now called distributed energy resources (DER). DER is becoming a major strategy in cities and regions across the globe (Green and Newman 2022) and is the basis of what we are calling global localisation.

Global localisation will mean global partnerships in development everywhere with a particular focus for each place that needs localised perspectives and localised smart technology to make it work. Some of the opportunities and options are set out below. The transport options that are emerging are all electric, integrated into the local sources of solar and wind, but there are also many ways of reducing the need for automobile dependence and also the need for travel by plane. The global economy has now demonstrated it can involve less travel as routine meetings are shifted to computer interactions that have worked well during COVID-19 (Earley and Newman 2021). However, for this to continue post-COVID-19 will need greater focus on localisation in centres that can uptake the new technologies. The re-localising of places that can enable people to value their place more (e.g. Beatley 2004) has been a growing social movement for the past 50 years and reached a peak during the COVID-19 lockdown (Allam and Jones 2020). Businesses will need to work simultaneously on interpreting global trends and creatively addressing local places where they want to invest. Place will be important, and each community will have new tools to make their places better through localised infrastructure management as well as localised services for a fuller range of activities.

Regenerative Ecosystems The idea of regenerative development, as explained above, has come from architecture and design where it has been the focus of individual buildings and groups of buildings that have sought to help repair the local environment. The new move is to take this idea and make it a much more global idea with application to broader environmental, social and economic assessment approaches including the zero carbon agenda (Newman 2019). The chapter will try to show how this new agenda of regenerative development will need to include how Smart Cities can play a critical role in its delivery.

The bigger development agenda story can be shown in Fig. 4.1, setting out how environmental and social impact of economic development has been addressed in three stages: exploitive development, sustainable development and regenerative development.

The three stages are as follows:

- First, environmental and social impact began to be addressed through direct regulations and assessment processes from the 1970s responding to highly exploitive development practices that had developed out of the modernist paradigm.
- Second, reducing economic and social impact shifted to the sustainable development paradigm as it was seen that resource issues like energy, water and climate change could not be addressed unless economic and social factors were fully integrated with the environmental agendas that had dominated the need to modify exploitive development; this began to flatten the curve but it was not transformative enough.
- Finally, we are shifting to regenerative development where the curve turns over and dramatically reduces impact until it is able to actually regenerate the environment (Thomson and Newman 2016). In climate terms, this will not just mean zero carbon but eventually extracting carbon from the atmosphere at a faster rate than it is going in (Flannery 2005). It will also enable zero poverty with a strong social agenda integrated into each element of development.

This is the hopeful approach that should be driving the next phase of economic activity, and there are signs that this regenerative approach may be an emerging paradigm as the need for a transformative outcome becomes more and more obvious (Newman 2020a) and more and more depends on cities showing how to do it

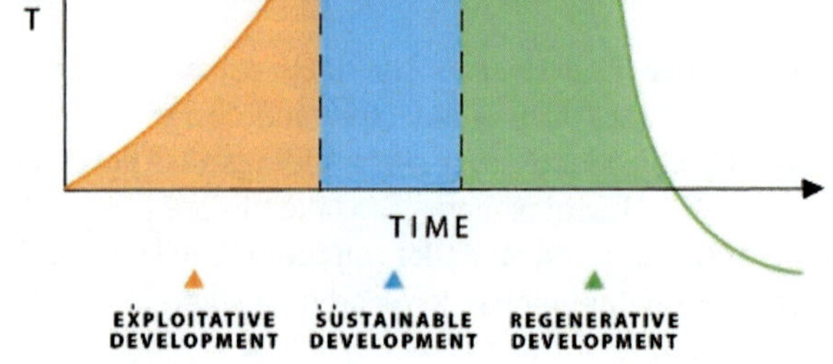

Fig. 4.1 The transition in approach to development from exploitive to sustainable to regenerative. (Source: Newman 2019)

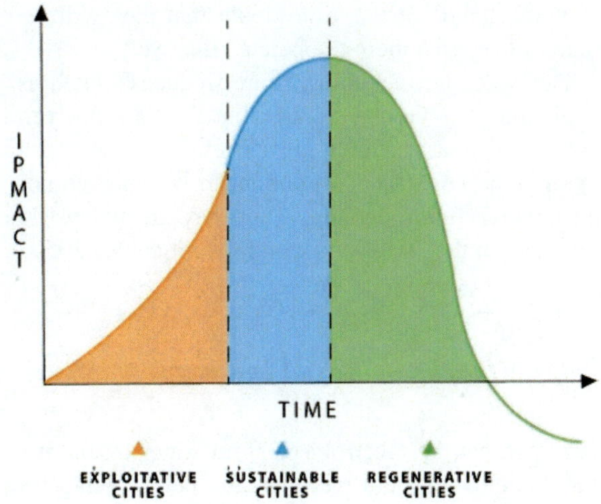

Fig. 4.2 Smart regenerative cities

(Thomson and Newman 2016, 2018, 2020). It is particularly apparent in the scenarios being presented by the IPCC in its mitigation approaches that suggest the need for repair of the atmosphere not just reductions in greenhouse emissions, in order to bring the planet below 1.5 °C of warming.

Regenerative ecosystem is the term that can be used to explain how the smart technology systems outlined in Chap. 3 can be used to help with this new and more challenging goal of regenerating the global and local environment as part of the global and local economic and social systems. This will need to work through Smart City systems.

Business, governments and communities will need to create the ecosystems of cities as neural networks which learn and grow showing us how to make each place in a city or region achieve regenerative outcomes. This is the regenerative ecosystem business model for development, and it will need to be explained in terms of how it can fit into different parts of the city along with globalised localisation. The next agenda is therefore smart, regenerative cities (Fig. 4.2).

4.3 The Immediate- and the Long-Term Technologies for Smart Regenerative Cities

4.3.1 The Immediate Term: Solar-EV-Batteries and Smart Technology Integration

As set out in Chap. 3, these technologies are dramatically growing and being adopted in many parts of the world and different parts of cities and regions to achieve zero carbon. They are mainstreaming as the flagship of the next industrial revolution, and there is little doubt that they will be using more and more smart systems to enable them to operate effectively.

The next four technologies are frequently cited as being part of this industrial revolution (e.g. Garnaut 2019), but they are not yet as commercial as the solar-battery-EV-smart technologies. They are at demonstration phase, and they could rapidly emerge, so it is important to try and see how they can fit into the smart regenerative future economy, but they are likely to be longer term yet still very important in the 2030s and certainly critical for 2050.

4.3.2 Hydrogen-Based Industry

Hydrogen can be electrolysed from water by the use of solar energy and can be stored chemically and mechanically before being used in a variety of ways. The hydrogen fuel cell can enable mobility, but at this stage, it is two to three times the cost of Li-ion battery mobility mostly because of the thermodynamics of having such a light and highly reactive substance like hydrogen. It is being completely left behind in the competition for land transport including with heavy trucks and freight trains as the hydrogen is so hard to store and distribute compared to electricity. The main potential for hydrogen in transport though seems to be for heavy, long-range ships and planes. Electric ships and planes are emerging rapidly for short distances and lightweight services (Whitehead et al. 2022). Green hydrogen can be converted into ammonia that can be fitted directly into diesel-fuel ships, and hydrogen can be combined with CO_2 from the atmosphere to create synthetic jet fuel (IPCC 2022).

The more rapidly emerging niche for hydrogen appears to be in industrial activity such as steel and cement production as well as metal refining (Garnaut 2019). Hydrogen is a rapidly growing commodity and should be part of the basis for next economy industrial growth as long as there is space for large solar-PV farms or wind farms and a water source to electrolyse for producing the hydrogen. The competitive advantage of different cities and their regions to be the next hydrogen-based industrial complex is already evident (Nahum 2020; West Australian Government 2019).

The need for smart systems with each part of the hydrogen economy is emerging whenever a new kind of demonstration has been developed. One example in remote area settlements where stand-alone power systems are required has emerged in Western Australia. The solar-battery systems being installed on farms and small towns have usually had a diesel generator as backup for the times when the battery is not enough and the sun is no longer shining enough. An innovation by a small firm called Blue Diamond Machinery working with a French innovation (https://www.bluedm.com.au/hydrogen-generators/) has created a hydrogen generator that uses a hydrogen fuel cell to provide the backup power; the hydrogen is created whenever the battery is filled and the excess power is used to hydrolyse water. The whole system is managed by smart systems that enable it to operate in all weather and is highly portable like diesel generators have been for decades. The potential to fit this into small microgrids in cities is also something that is on the agenda.

4.3.3 *Circular Economy*

Cities have been innovating for some time to reduce their metabolism, i.e., their resource inputs and their waste outputs (Newman and Kenworthy 1999). This has been given a new look as the agenda for a circular economy (Dohmen and Confiado 2018; Petit-Boix and Leipold 2018). The technology for waste disposal in the past has been centralised, large scale and largely linear not circular, i.e. it has had little emphasis on recycling unless cities have been running out of space. The new systems for the circular economy are, like the other innovations discussed above, much smaller in scale and are able to be used in more localised and distributed situations, even in slum areas (Teferi and Newman 2018). Again, this small-scale approach to decarbonising the future will always need smart systems to integrate them into the cities where circular economy options are being sought.

An example of circular economy integration with smart systems is how bottles are recycled now by putting them in a reverse vending machine that can dispense money based on a sensor checking the bottle type or how lithium-ion battery metals (Li, V, Mn, Co, Ni) can be traced from their origins after being processed by doping and having blockchain trace exactly where they are and when they reach a recycling centre showing the industry facility what are their exact element compositions, so they can be sorted and extracted for future use.

4.3.4 *Biophilic Urbanism, Permaculture and Nature-Based Solutions*

The modernism that helped build cities after the 1930s collapse did not have a strong awareness of the natural environment, or at least how we see it today. It is 50 years since the first Earth Day showed how the blue planet viewed from space was facing new limits now described as the Anthropocene's planetary boundaries (Steffen et al. 2015; Meyer and Newman 2020). The biggest part of that agenda is global climate change from fossil fuels and land clearing, which the above technological approaches are meant to address along with the other SDGs. But the local natural environment agenda within cities have also been growing and have now coalesced into two areas – biophilic urbanism and permaculture.

Biophilia was put on the agenda for cities by E. O. Wilson who showed that we cannot put ourselves above nature in our cities as we coevolved and need nature in our daily lives (Wilson 1984). The response has been a set of technological approaches that has shown how cities can build natural systems into and onto buildings with green roofs and green walls and which converts water systems from drains to natural water courses (Beatley 2016; Soderlund 2019; Keller et al. 2008). The best examples have been in dense cities like Singapore that have been able to use high-rise structures as habitat like in forests (Newman 2014; Newman et al. 2012). The Biophilic Cities Network is growing across the globe and found that local biophilic features played a very strong role in the COVID-19 lockdown providing a healthy link to nature (https://www.biophiliccities.org/COVID-1919-research). Biophilic urbanism is likely to be a feature of many new urban developments as the cost-effective attractions and local community building potential are high.

Biophilic urbanism will always be available as a simple set of landscaped areas, but to make high-intensity biophilics work in dense cities, there will need to be smart system integration. For example, in Singapore, the intensely developed green walls are fed with irrigation water dosed with fertilisers that are automatically added depending on what dosage is signalled by sensors in the soil.

Permaculture is similar to biophilic urbanism in its desire to link cities and nature using new kinds of technology and different approaches to how to make urban ecosystems work better (Mollison and Holmgren 1978) but was essentially developed for suburbs with large spaces around houses, rather than the density where much biophilic urbanism has been applied. The new version of permaculture developed by Holmgren (2020) is called RetroSuburbia and is designed to rebuild suburbs around a new vision of shared economy, food growing and re-localised services. It may emerge with considerably more intensity as middle and outer suburbs begin to be rebuilt.

The idea of smart systems being needed to make permaculture more effective is not part of the history of this deeply held paradigm for how to shape the future. However, the principles of permaculture can be applied to any settlement and any kind of urban fabric no matter how dense, so it will be possible to find emerging examples of how this can be integrated into the future zero carbon-zero poverty city.

Some examples include how the principle of enabling ecological mixes of plants in gardens can be designed to attract insects and birds for year-round biodiversity attractions, but the actual activity may need visual sensors that can be recorded to show bird activity. Water sensitivity will always be sensitive no matter what kind of permaculture and what climate, but the value of sensors that can indicate when water stress is occurring will always be of value in a city where climate and soils will vary a lot.

The two concepts can be enriched in Smart Cities by ensuring they are intertwined with emerging trends, such as the adoption of nature-based solutions (NbS) within the urban sphere. These solutions seek to not only enhance the greening of cities but also ensure that nature is both preserved and to be maximally utilised to guarantee food sovereignty and reduce the impacts of unpredictable weather conditions (e.g. excessive flooding due to high precipitations) (Wendling et al. 2018). Additionally, the quality of life is expected to be improved via better air quality, increased recreation spaces and green infrastructures such as tree-covered bicycle lanes, among others. An array of solutions now exists. For example, nature-based solutions such as permeable pavements have the capacity to reduce the impacts of flooding by allowing water runoffs to penetrate the open spaces within the pavements (Muttuvelu et al. 2022). This is part of sustainable drainage systems that could be adopted in cities across the world, making them resilient against the impacts of climate change. In addition, excessive runoff waters could be redirected into artificial rain gardens that not only help reduce flooding but have the potential to increase the food supply in cities, thus guaranteeing food security.

Another nature-based solution that could be integrated into the Smart City agenda is the creation of man-made, urban wetlands. These are argued to hold immense potential in sustainability agendas as they serve as wastewater filters, hence helping in ensuring a sustainable supply of clean water in cities. Wetlands also serve as buffer zones for water bodies, mainly from various pollutants that are on the rise in urban areas as the population in cities continue to increase (Rogerson et al. 2021). In addition to protecting water bodies, wetlands have the potential to provide habitat for wildlife, help regulate urban temperatures and open opportunities for economic activities like aquaculture (Cross 2018). The use of smart technologies such as high-resolution satellites would be helpful in identifying areas within cities that could be most potent for constructing the wetlands, therefore allowing for the optimal utilisation of available land and resources and promoting other sustainable practices like conservation (Mahdianpari et al. 2020).

Adopting emerging urban planning models, especially those that promote greener and healthier mobility options, such as bicycles and walking, needs to be considered as key elements for urban regeneration. The impacts of conventional transportation options on the environment, human health, the economy and natural resources have been widely documented. There is a universal agreement on the need to shift to more sustainable, cost-effective, greener and economically accessible options. Cycling and walking have been fronted as being the most optimal for most cities. Experimentation with these options in cities like Paris has proven that these mobility options are practical and achievable. The advantage of these options is that they can utilise the existing transportation corridors, provided the lanes are strictly dedicated to alleviating negative impacts such as accidents. In doing so, further addition of hard infrastructure and mega projects can be reconsidered, in favour of softer options, which can also render more compact, diverse and vibrant cities, living in harmony with both people and nature.

4.3.5 Fitting These New Technologies into a Smart Regenerative City Through Local Community

The agenda to create solar-battery-electromobility-Smart City developments are now rolling out into all parts of cities and regions and are trialling different smart technology systems in different places. As these systems become established, it is likely that they will begin to incorporate circular economy, biophilic urbanism and permaculture as they are all complementary to the decarbonised power agenda that is driving solar-battery EVs. All of these technological systems need to focus on a re-localised place in order to provide the Smart City integration necessary to make the system commercial and hence mainstream. These integrated systems would all be able to use AI or machine learning as the connections can be optimised as a system rather than as separate systems. It means that cities and settlements with these smart regenerative systems would be an ecosystem with the equivalent of neural networks that enable it to respond to the changing weather and changing demands. This need not be expensive as smartphones can be used to manage them.

There will be a multiple set of agendas pushing this re-localising smart regenerative city process, especially if it has the neural networks of machine learning that can be used for addressing adaptation to climate emergencies or future pandemics. But the mainstreaming of zero carbon and zero poverty will continue to be the main issue. All of the agendas depend on mobilising the social capital of a place and build the resilience that will be increasingly needed. In a film made through Curtin University and Canterbury University called *Christchurch: Resilient City*, the theme was how the rapid recovery of the city after one of the worst earthquakes in history was based on their strong local communities with their social capital able to help save lives (Blagg et al. 2014).

Such social capital forms the basis not just for responding to disasters but to creating long-term economic change. Social capital is critical for the localised technologies needed for zero carbon futures but also new job creation activity with local entrepreneurs to create zero poverty outcomes. Local social capital provides the support base needed to create new businesses as well as the new support systems provided by multiple smart, regenerative businesses all seeking to achieve the global outcomes of zero carbon and zero poverty.

Such local social capital is not just found in developed cities but is probably more evident in emerging cities where significant slum housing is evident. The approach to slum housing in the past has often been to clear the sites and move people into high-rise housing. Such modernist solutions have constantly failed as the tight social capital that helped hold the slum communities together is immediately lost. In a detailed study of slums in Jakarta and Addis Abba, detailed studies of people in slums were compared to those moved from slums into high-rise housing; the slum communities had significantly greater trust in their neighbours, shared more and provided money and help significantly more (Asseffa and Newman). The conclusions were made easier in suggesting that the policy approach should be to use new zero carbon technologies to enable better power, water and sewerage, as these are now small scale and fit into such places effectively and cheaply without destroying the social capital (Asseffa and Newman). The new technology can indeed enable the socio-technical breakthrough for slum housing in creating the opportunity for the place to be given formal status instead of being informal that makes residents unable to have an address with access to banks and formal workforce opportunities. The smart systems of sharing can all be run through mobile phone apps that most in slums can use. Such a transition should be part of the acceleration into the future Smart Cities' agenda.

4.4 Planning Implications for Smart Regenerative Cities

4.4.1 *Create Re-localised Centres with Smart City Regenerative Outcomes*

Planners and other urban professionals have a new focus as they can implement all the major technological approaches for driving the new economy through *re-localising centres*. The hydrogen industry agenda is slower than solar-battery EVs but will probably come within a decade as the technology becomes more commercial. An example of an early hydrogen technology being used in remote areas is set out below. The other emerging technologies are circular economy, biophilic urbanism and permaculture which are all being trialled in various demonstrations but are not yet mainstreamed.

Re-localised centres will be very different places depending on the fabric in which they will be built (as outlined below). The key issue for planners is to ensure that common good outcomes are set for each localised centre. These can become regenerative outcomes if Smart City systems are used to learn how this can be done in ways that locals can envisage as well as ensuring that they are dense enough to provide the necessary services and that they are provided with all the required infrastructure, especially transport (Newman 2016). These needs are best understood if the different qualities of urban fabrics are seen as the basis for these new smart regenerative cities.

4.4.2 *Tailor Re-localised Centres for Each Urban Fabric*

One of the big problems with the modernism created by Le Corbusier and the Athens Charter in 1933 was that the modernism movement saw 'one best way' to create the city of the future. There was one urban system that needed to be modernised with freeways and housing estates often used as battering rams to 'clean the slate' (Jacobs 1961). Historical battles resulted and continued up until recently (Gaynor et al. 2017). These responses across the world have shown that walking urban fabric in the historic central area and the transit corridors created by the train and tram eras, as well as the new automobile-based suburbs outside these areas, are significantly different in their urban qualities and hence in how they are managed. As outlined in Chap. 3, the theory of urban fabrics shows how these three fabrics should be recognised, respected and rejuvenated differently (Newman et al. 2016). A further two fabrics were added due to the differences observed in the peri-urban fringe areas of cities which merge into rural village fabric (Newman et al. 2017), as well as remote settlements where mining camps, indigenous settlements and other smaller villages (often temporary recreational sites) are located (Newman et al. 2019). Each of these needs solutions to enable them to create re-localised centres tailored to use their fabric for the best contributions to the smart regenerative cities of the future.

In Table 4.1, the core technological approaches that were outlined above are listed down in the left column, and the five different types of urban fabric are spread across the top. Inside the boxes are ticks that show the extent to which the core technological approach applies to that fabric.

4.4.2.1 Applying Smart Regenerative Technologies to Urban Fabrics

1. Central city walking cities are less able to install solar-PV but are ideal for walkable active transport and micromobility, smart systems, as well as biophilic urbanism. Circular economy and permaculture don't work here without significant extra costs, but this may change as technologies improve for particular functions that fit high-rise buildings. The changes that enable solar to be part of glass and recycled wood in CLT are already being designed into high-rise buildings. The role of smart systems in cities for surveillance and information in smart and shared transport systems shows that the neural network concept may well emerge in old central cities as they move into the sixth wave of innovation after seeing out all five before this. They are usually very strong in the social capital needed to enable them to adopt to change, so we can expect a range of demonstration sites in central cities to emerge in the next decade.
2. Transit city corridors are better for solar-PV and batteries, ideal for transit, micromobility and active transport, with some potential circular economy and biophilics with permaculture possibilities (perhaps in community spaces). These areas have generally been regenerated in the past decade as the move back into less automobile-dependent urban fabrics accelerated (Newman and Kenworthy 2015a). As they have a higher concentration of housing than central cities with their emphasis on offices, these areas have developed strong commitments to new technology businesses that are able to utilise the urban fabric for face-to-face activity such as project development and management operations. They are also where many education and health facilities exist. Such urban fabric is ideal for enabling smart regenerative demonstrations such as zero carbon corridors (Newman et al. 2021). The station precincts that already exist can be transformed into microgrid-based solar systems and many of the other technologies enabled to feed into such grids providing the basis for zero carbon products and services. Such areas are likely to grow rapidly if enabled to function in this way.
3. The middle and outer suburbs of the automobile era are very good for solar-PV (in Australian cities, most of the poorer outer suburbs installed PV first (Newton and Newman 2013)), circular economy and permaculture but will need EV cars and buses with some new zero carbon corridors helping overcome automobile dependence. The industry estates could be enabled with hydrogen-based services for industrial processing as well as solar. These would all be larger-scale projects as the space is greater, thus enabling processing and manufacturing of circular economy materials into a large array of zero carbon products.

Community-based permaculture can become the basis of social capital for enabling those who live in the area to be part of cooperatives for such green produce. All of these activities will need smart systems but at larger geographical scales and for different types of production and services.

4. Rural villages and peri-urban areas will need to form new localised centres in order to make the most of the benefits of power and transport with integrated solar-PV-battery-electromobility and progressively with electrified agricultural vehicles feeding into stand-alone systems on farms. In these areas, permaculture food production has big opportunities as well as circular economy jobs and hydrogen-based industry at the scale that is presently seen in heavy industrial sites. All of these new technologies will again require smart systems, and they will have localised functions but at a bigger scale than in the city as they may not have any grid to depend on other than their own microgrid.
5. Remote areas are ideal for microgrids of integrated solar-PV-batteries and electromobility whether in small indigenous villages or mining camps. In addition to these living sites that will generally have stand-alone power, there is a need to provide for the zero carbon future with very large-scale solar and hydrogen-based industries. Already vast areas are being set aside for solar-PV farms that can provide the basis for the power and hydrogen-based industry that would be established mostly adjacent to mining areas. They will need plenty of space for substantial solar and wind farms for hydrogen production. This hydrogen is much better used locally as it is a very hard product to store and distribute but is very effective if used locally. Thermodynamics will favour localised green hydrogen production and usage in industrial processing of minerals and even agricultural products (Whitehead et al. 2022). As outlined above, there are already commercial small-scale hydrogen generators that are being fitted to stand-alone power plants that presently have solar, batteries and backup diesel generators https://www.bluedm.com.au/hydrogen-generators/. Needless to say, this new technology needs smart systems to enable it to work, and it is being rolled out in the rural and remote areas of Australia for mining camps, remote indigenous settlements and farmers who are at end of grid and need more reliable systems. It is a very good example of good global localisation as it is likely to be highly relevant to any village or remote settlement and enable leapfrog economic development (Ndlovu et al. 2021).

4.4.3 Reorient Transport and Urban Development Priorities to Suit Fabrics

The re-localised centres will be made viable or not by the transport priorities chosen to drive the urban regeneration. As Table 4.1 shows, there is a need to respect and regenerate each fabric type. For example:

1. The central city needs to be a place for walking as that was its fabric and the demand to continue this is obviously growing. Models for this have been created around the world (Gehl 2010). Pedestrian and cycling infrastructure are the highest priority following the lead of London during COVID-19 (Quinn 2020).
2. The transit city needs to be a place for high-quality transit and urban regeneration around its stations. The demand for more of this has created the notion of transit activated corridors along main roads where new mid-tier battery-electric transit can move quickly along corridors and slowly through regenerated centres around stations where micromobility can feed the service and link the surroundings to their local activity centre (Newman et al. 2018). Making space in the road for such high capacity transit will be the next big issue in transport prioritising politics as main roads now need to be seen as major transit corridors with an important role in creating net zero urban regeneration.
3. The automobile city needs to be where electric cars are at home as well as micromobility for linking to new re-localised shopping and services centres but also to jobs in industrial estates, circular economy/recycling centres and permaculture food-growing areas. Finding ways to enable local car journeys but not cross city journeys will be a critical disjunction with fourth wave modernist cities. This area can include autonomous electric vehicles, but they are unlikely to be part of walking and transit fabric (Newman et al. 2019).
4. Edge city fabric or peri-urban and rural villages need to have a multitude of functions but mostly with a strong local focus; electric micromobility can still work in these areas as it can link to such areas for a 10 km radius without being too slow. Electro-mobility is likely to grow rapidly for such areas, e-micro-mobility, EV's and for agricultural purposes electric tractors and farm machinery. However, they will still need to grow more compactly in rural villages to enable opportunities for the localised services to be provided, which could include an electric bus service to help those without electric vehicles. Models such as Witchcliffe Ecovillage can easily become the norm as they are cost-effective and on the right side of history in terms of climate-resilient development.
5. Remote villages for indigenous and mining functions need to have electric off-road vehicles linked to their solar-based recharge hubs. Many examples of solar villages have been demonstrated in both kinds of settlements, and it is now time to mainstream these.

The density of the re-localised 'town' centres in each fabric will reduce from the centre to the urban fringe but will need to be compact and walkable, or they will not work even further out in rural and remote areas (Newman et al. 2016; Thomson et al. 2016). Many mining villages are like medium density urban villages. Solar and battery recharge hub services will need to be available in all these re-localised centres as well as delivery hub services and as much Smart City support services to enable it to work as zero carbon places while enabling economic activity to flourish.

The developing world will need to approach each of these opportunities as though it was their chance to leap frog into the future as already economic development based around solar-PV-battery-electromobility-Smart City innovations is challenging traditional development as they are more appropriately scaled and have multiple benefits (Intergovernmental Panel on Climate Change (IPCC) 2019). China has led the way in many of these innovations being mainstreamed, and it is showing in their cities (Gao and Newman 2018). The innovations, being small scale, are also much more relevant to the vast areas of slum developments in the developing world that in modernism were destined only to be cleaned out, thus losing much of their important community structures (Teferi and Newman 2018). As outlined above, these tight structures are ideal for the small-scale energy, water and waste infrastructure opportunities that can be managed locally (Teferi and Newman 2017). Mobile phones have rapidly become part of slums, so in a similar rapid process, developing cities could have better infrastructure in the post-COVID-19 recovery for both zero carbon and zero poverty. The same pattern across the great megacities of the developing world can pick up on the local community structures so common to them all and enable regenerative outcomes to be installed through these new technology systems (UNCRD 2019). Most of all, it will be zero carbon with all the SDGs including a strong local employment generator.

4.5 Partnerships to Fund the Smart, Regenerative City

Re-localised centres and transit activated corridors using new technology systems all need to be funded. Public-private partnerships (PPPs) are the ideal approach for this integrated set of urban redevelopment technologies. This is not only because capital is scarce for governments but because the best way to create projects that are viable is around value uplift. Value is created from transit infrastructure, walkability, biophilic features and other urban regeneration features (McIntosh et al. 2015; Matan and Newman 2016; Cabanek and Newman 2016). This can probably now be extended to all kinds of new re-localised centres across the city such as industrial estates, permaculture villages and rural and remote villages.

The package of technologies is symbiotic, and so is the need for symbiotic partnerships between all levels of government, the private development industry and local communities. The advantage in creating symbiotic partnerships is that in the process of using community-based approaches (Hartz-Karp and Marinova 2012), long-term commitments are enabled which form the basis for good government decisions, good financing arrangements and good community. Community values are fundamental to all viable, flourishing cities (Newman 2018). With strong community values supporting zero carbon-zero poverty development, it is likely to be much more feasible. Shared economy demonstrations with symbiotic partnerships are growing in number, and most are based around zero carbon-zero poverty values (Sharp 2018; Salter et al. 2019).

Cities cannot achieve the zero carbon-zero poverty agenda without substantial redevelopment, especially in their declining middle suburbs where old car-dependent suburbs are not finding the right model for redevelopment (Newton et al. 2012). As outlined above, the notion of transit activated corridors could enable this process if set up with symbiotic partnerships to enable urban regeneration for the new economy. This is likely to be the biggest focus of development as the middle suburbs are the area where housing has reached the end of its life. Redevelopment of the 'missing middle' with the zero carbon-zero poverty agenda should involve a substantial proportion of affordable, social housing.

4.6 Rewrite Manuals for the Smart Regenerative City

The manuals of modernism are alive and well on most urban professionals' desks or at least in their computer models and assessment procedures based on an idea from the modernist tradition that there is a 'one best way' that can be applied across all parts of cities. The need to seek common good outcomes in planning and urban practices in general will still be needed, but there needs to be a new set of manuals created to go with the new economy, especially as they need to apply different approaches to each of the five urban fabrics (Newman 2016, 2017). There needs to be new zero carbon-zero poverty manuals for planning, transport, energy, water and waste, in fact in every area of the economy developed since the 1940s, with a focus on how they can be regulated and prioritised in their infrastructure, density and functions in a far more nuanced way. They will also need to understand how the integration of smart systems creates a whole range of new opportunities which undermine the need for simplistic modernist regulations that fit every part of the city with the same requirements. The new manuals will be much more focused on outcomes and demonstrations of best practice.

This will not be easy as the socio-technical transition literature shows that there are endless barriers already preventing the major innovations being demonstrated for decarbonising the economy (Geels Frank et al. 2017). This can be seen especially in the culture of professionals and what they must do to show that their work matters. Rediscovering what matters and creating work around that will be an important new task for urban professionals everywhere. Perhaps the triage thinking of COVID-19 professionals can help work through new approaches to assessing projects so that projects don't die due to years of linear, modernist regulatory assessments that are no longer relevant (Newman 2020b).

4.7 The Human Dimension: Cultural Urban Renaissance

John Montgomery (2017) followed up the theory of waves of innovations that changed economies across the globe in the last 200 years, by adding a human dimension – how cultures are part of this change. In particular, they are part of how they make cities come alive with different cultural expressions and provide the underlying values that enable the city to gain a new hope for the future underlying such activity.

What kind of cultural dimension can underlie this change toward re-localised cities with reduced automobile dependence? It will more than likely be more urban not less urban. The fourth wave based on cars and oil was a suburban era, and there was a decline in inner and central cities as cities spread out following suburbanised work, housing and shopping (Newman and Kenworthy 1989, 1999, 2015a). However, the ICT-based fifth wave which suggested that cities may not even be needed as we could just scatter and telecommute (Webber 1963, 1964, 1968) did not happen, but instead, it was the basis of a new revival of urbanism that was based on the knowledge economy (Glaeser 2011). This required face-to-face meetings for the creative development of projects and soon created the phenomenon of peak car and the millennial generation movement back into more urban locations to live and work (Newman and Kenworthy 2015b). Florida (2019) has now suggested this was not deep enough in its ability to shape the city and rapidly ran out of the urban fabric that could be revived. However, the idea of re-localised centres in all five urban fabrics provides the basis for new, more urban development with a cultural basis in each of the five areas that supports its diversity of forms.

The human dimension of a city will be driven by the extent to which the formal and informal creative sector is able to help shape urban development; the evidence of recent decades would suggest that this will be with a more urban flavour and focus. This will be different in different cultures, but without a strong urbanism driving the future, there will not be a fundamental move away from the heavily car-dependent fossil fuel-based city of the mid to late twentieth century for which much of the creative sector are now looking for something more (Gehl 2010). Florida (2018) suggests that the creative industry is a key contributor to urban economic growth. Now we sit poised with all the innovation tools that can enable a more robust urban culture and a more equitable and sustainable city, so we just need to add the creative industry with their urban flavour. Perhaps this city's time has arrived, more so riding on emerging modern technologies such as AI, digital twins and 4G and 5G that have become key components of the creative industry.

4.8 Conclusions

The growth of cities has been the basis of civilisation (Hall 2008), and thus cities will bring together the integrated agenda of zero carbon and zero poverty in a new phase of civility. Smart, regenerative cities in the future will simultaneously be creating a better global and local environment while creating liveability through a better economy and increased social interactions. Some have seen a collapse of civilisation (Diamond 2005) and even 'civicide' (Hancock 2019) as the only realistic outcome from the pathways that our cities were following – until the COVID-19 collapse (Newman 2020b). Now we have the opportunity to create a new kind of civility based on zero carbon-zero poverty and focused attention on inclusivity and sustainability, which will need the following technological approaches:

- Renewable energy especially locally applied solar-PV.
- Batteries and storage systems that enable renewables and electromobility.
- Electromobility and associated new transit, walkability and micromobility.
- Hydrogen-based industry replacing the last of the big systems needing fossil fuels: food processing, steel and other mineral processing (especially all the battery minerals) and cement production.
- Circular economy systems applied in industrial estates and through local communities in each urban fabric.
- Biophilic urbanism that brings natural systems on and inside buildings to achieve new urban habitats and green infrastructure.
- Permaculture that can add new ways of growing food in the various different fabrics of cities and regions.
- Smart city technology that integrates all the above and enables all of these innovations to work better. It is a critical feature and creates the ecosystems of cities as neural networks which learn and grow showing us how to make each place in a city or region achieve regenerative outcomes.

The first group of changes – renewables, batteries and EVs – appears to be already cost-effective as they are showing their prowess at supporting economies in the immediate post-COVID-19 rise to zero carbon economic recovery. The hydrogen, circular economy biophilics and permaculture are rapidly moving to be cost-effective as they move up the multilevel perspective transition from niches to regimes of demonstrations (Geels Frank et al. 2017).

But the most significant point made in this chapter is that smart technology and systems are needed to integrate these technologies into each part and each fabric of the urban and regional systems that drive our economy. Investment will need to have assessment and planning processes that enable rather than drive away such changes and will need new manuals and new partnership processes that enable all levels of government to work with private finance and business, in close collaboration with communities whose place-based values need to drive all development and help create local enterprises.

These processes will be a challenge for all urban leaders and professionals who must do more than simply wait for such technologies to solve everything. The need will be for solutions that combine these technologies with regenerative ecosystems applicable to particular places: in dense city centres, inner city corridors, suburban centres with estates for various functions and peri-urban, rural and remote villages. The human qualities of each place will be at the heart of how these new technologies can create re-localised places.

Urban professionals will need to rapidly change the manuals of modernism still so prevalent in their fourth wave engineering designs and statutory regulations. Cities that can quickly focus on how to mainstream their new planning and assessment systems to create new centres of zero carbon-zero poverty urbanism are likely to be the new centres of civilisation, especially if they can work synchronously with the creative side of their cities.

Smart cities need to be redefined into the regenerative design model of the future, or it will never have the values that determine the political drivers of change.

References

Allam Z, Jones DS (2020) On the coronavirus (COVID-19) outbreak and the smart city network: universal data sharing standards coupled with artificial intelligence (AI) to benefit urban health monitoring and management. Healthcare. Multidisciplinary Digital Publishing Institute 8:46

Beatley T (2004) Native to nowhere: sustaining home and community in a global age. Island Press, Washington, DC

Beatley T (2016) Handbook of biophilic city planning & design. Island Press, Washington, DC

Blagg L, Beatley T, Newman P (2014) Christchurch: resilient city. Vimeo

Cabanek A, Newman P (2016) Biophilic urban regeneration: can biophilics be a land value capture mechanism? In: Brebbia CA, Zubir SS, Hassan AS (eds) WIT transactions on ecology and the environment. Sustainable development and planning. WIT Press, Southampton, pp 65–78

Cross K (2018) Wetlands can provide a pathway for a sustainable urban future. Available at https://iwa-network.org/wetlands-can-provide-a-pathway-for-a-sustainable-urban-future/#:~:text=Urban%20wetlands%20contribute%20to%20the,development%20for%20a%20sustainable%20future. Accessed 10 Dec 2022

Diamond J (2005) Collapse: how societies choose to fail or succeed. Viking, New York

Dohmen G, Confiado A (2018) Circular economy indicators: what do they measure? Availabe online https://www.unep.org/news-and-stories/blogpost/circular-economy-indicators-what-do-they-measure. Accessed 8 July 2020

Earley R, Newman P (2021) Transport in the aftermath of COVID-19-19: lessons learned and future directions. J Transp Technol 11:109–127. https://doi.org/10.4236/jtts.2021.112007

Flannery T (2005) Atmosphere of hope: solutions to the climate crisis. Text Publishing Company, Melborne

Florida R (2018) How culture shapes economic development. City Lab

Florida R (2019) The rise of the creative class. Basic Books, New York

Gao Y, Newman P (2018) Beijing's peak car transition: hope for emerging cities in the 1.5 C agenda. Urban Plan 3:82–93

Garnaut R (2019) Superpower: Australia's low-carbon opportunity. Schwartz Books Pvt Limited, Victoria

Gaynor A, Newman P, Jennings P (2017) Never again: reflections on environmental responsibility after Roe 8. UWA Scholar Press, Perth

Geels Frank W, Sovacool Benjamin K, Schwanen T, Sorrell S (2017) Sociotechnical transitions for deep decarbonization. Science 357:1242–1244. https://doi.org/10.1126/science.aao3760

Gehl J (2010) Cities for people. Island Press, Washington, DC

Glaeser E (2011) Triumph of the city: how our greatest invention makes us richer, smarter, greener, healthier, and happier. Penguin Publishing Group, London

Green J, Newman P (2022) Transactive electricity markets: transactive electricity: how decentralized renewable power can create security, resilience and decarbonization. In: Droege P (ed) Intelligent environments 2 – advanced systems for a healthy planet. Elsevier, Berlin

Hall P (2008) Cities and civilization. Pantheon Books, New York

Hancock PA (2019) In praise of civicide. Sustain Earth 2:8. https://doi.org/10.1186/s42055-019-0014-9

Hartz-Karp J, Marinova D (2012) Methods for sustainability research. Edward Elgar, London

Hes D, Du Plessis C (2014) Designing for hope: pathways to regenerative sustainability. Taylor & Francis, New York

Holmgren D (2020) Retrosuburbia: the Downshifter's guide to a resilient future. Melliodara, Hobart

Intergovernmental Panel on Climate Change (IPCC) (2019) Global warming of 1.5°C. An IPCC special report on the impacts of global warming of 1.5°C above pre-industrial levels and related global greenhouse gas emission pathways, in the context of strengthening the global response to the threat of climate change, sustainable development, and efforts to eradicate poverty. World Meteorological Organization, Geneva

IPCC (2022) Climate change 2022: mitigation for climate change. IPCC

Jacobs J (1961) The death and life of great American cities. Vintage Books, New York

Keller R, Heerwagon J, Mador ML (2008) Biophillic design: the theory, science and practice of bringing buildings to life. Wiley, Hoboken

Mahdianpari M, Granger J, Mohammadimanesh F et al (2020) Smart solutions for smart cities: urban wetland mapping using very-high resolution satellite imagery and airborne LiDAR data in the City of St. John's, NL, Canada. Remote Sens 280(1):111676

Matan A, Newman P (2016) People cities: the life and legacy of Jan Gehl. Island Press, Washington, DC

McIntosh JR, Newman P, Trubka R, Kenworthy J (2015) Framework for land value capture from investments in transit in car-dependent cities. J Transp Land Use 10. https://doi.org/10.5198/jtlu.2015.531

Meyer K, Newman P (2020) Planetary accounting. Springer, Berlin/Heidelberg

Mollison B, Holmgren D (1978) Permaculture one: a perennial agriculture for human settlements. Transworld Publishers, Melbourne

Montgomery J (2017) The new wealth of cities: city dynamics and the fifth wave. Routledge, London

Muttuvelu DV, Wyke S, Vollertsen J (2022) Are permeable pavements a sustainable solution? A qualitative study of the usage of permeable pavements. Sustainability 14(19):12432

Nahum D (2020) Powering onwards: Australia's opportunity to reinvigorate manufacturing through renewable energy. The Australia Institute Centre for Future Work/ACT

Ndlovu V, Newman P, Sidambe M (2021) Prioritisation and localisation of sustainable development goals (SDGs): challenges and opportunities for Bulawayo. J Sustain Dev 13:104. https://doi.org/10.5539/jsd.v13n5p104

Newman P (2014) Biophilic urbanism: a case study on Singapore. Aust Plan 51:47–65. https://doi.org/10.1080/07293682.2013.790832

Newman P (2016) Sustainable urbanization: four stages of infrastructure planning and progress. J Sustain Urban Plan Prog 1:3–10. https://doi.org/10.18063/JSUPP.2016.01.005

Newman P (2017) The rise and rise of renewable cities. Renew Energy Environ Sustain 2. https://doi.org/10.1051/rees/2017008

Newman P (2018) The theology of sustainability practice. In: Marques J (ed) Handbook of engaged sustainability. Springer, Cham

Newman P (2019) Driverless vehicles and pedestrians don't mix. So how do we re-arrange our cities? Availabe online https://theconversation.com/driverless-vehicles-and-pedestrians-dont-mix-so-how-do-we-re-arrange-our-cities-126111#:~:text=Cities%20mostly%20are%20planning%20more,point%20to%20enable%20vehicle%20access. Accessed 8 July 2020

Newman P (2020a) Cool planning: how urban planning can mainstream responses to climate change. Cities 103:102651. https://doi.org/10.1016/j.cities.2020.102651

Newman P (2020b) COVID-19, cities and climate: historical precedents and potential transitions for the new economy. Urban Sci 4. https://doi.org/10.3390/urbansci4030032

Newman PWG, Kenworthy JR (1989) Gasoline consumption and cities. J Am Plan Assoc 55:24–37. https://doi.org/10.1080/01944368908975398

Newman P, Kenworthy J (1999) Costs of automobile dependence: global survey of cities. Transp Res Rec 1670:17–26. https://doi.org/10.3141/1670-04

Newman P, Kenworthy J (2015a) Automobility and resilience: a global perspective. In: The urban political economy and ecology of automobility driving cities, driving inequality, driving politics walks. Routledge, pp 28–58

Newman P, Kenworthy J (2015b) The end of automobile dependence: how cities are moving beyond car-based planning. Island Press, Washington, DC, pp 1–300. https://doi.org/10.5822/978-1-61091-613-4

Newman P, Blagg L, Beatley T (2012) Singapore: biophilic city. YouTube. https://youtu.be/XMWOu9xIM_k

Newman P, Kosonen L, Kenworthy J (2016) Theory of urban fabrics: planning the walking, transit/public transport and automobile/motor car cities for reduced car dependency. Town Plan Rev 87:429–458. https://doi.org/10.3828/tpr.2016.28

Newman P, Beatley T, Boyer H (2017) Resilient cities, second edition: overcoming fossil fuel dependence. Island Press, Washington, DC

Newman P, Wills R, Edwards C, Yates C (2018) Lithium valley: building the case to establish Western Australia as a manufacturer of energy metals and power storage. Regional Development Australia, Kwinana Industries Council and Curtin University, Perth

Newman P, Hargroves K, Davies-Slate S, Conley D, Verschuer M, Mouritz M, Yangka D (2019) The trackless tram: is it the transit and city shaping catalyst we have been waiting for? J Transp Technol 9:31–55. https://doi.org/10.4236/jtts.2019.91003

Newman P, Davies-Slate S, Conley D, Hargroves K, Mouritz M (2021) From TOD to TAC: why and how transport and urban policy needs to shift to regenerating main road corridors with new transit systems. Urban Sci 5:52. https://doi.org/10.3390/urbansci5030052

Newton P, Newman P (2013) The geography of solar PV and a new low carbon urban transition theory. Sustainability 5:2537–2556

Newton P, Newman P, Glackin S, Trubka R (2012) Greening the greyfields. Int J Soc Behav Educ Econ Business Ind Eng 6:2870–2889

Petit-Boix A, Leipold S (2018) Circular economy in cities: reviewing how environmental research aligns with local practices. J Clean Prod 195:1270–1281. https://doi.org/10.1016/j.jclepro.2018.05.281

Quinn B (2020) London cycling could increase tenfold after lockdown. Availabe online https://www.newsbreak.com/news/0OyjFBww/london-cycling-could-increase-tenfold-after-lockdown-says-tf. Accessed 8 July 2020

Rogerson RJ, Horgan D, Roberts JJ (2021) Integrating artificial urban wetlands into communities: a pathway to carbon zero? Front Built Environ 7:777383

Salter R, Merson J, Rauland V, Odell P, Sharp D (2019) Engaging local communities. In: Newton P, Prasad D, Sproul A, White S (eds) Decarbonising the built environment: charting the transition. Palgrave Macmillan, Singapore

Sharp D (2018) Sharing cities for urban transformation: narrative, policy and practice. Urban Policy Res 36:513–526. https://doi.org/10.1080/08111146.2017.1421533

Soderlund J (2019) The emergence of biophilic design. Springer, London

Steffen W, Richardson K, Rockström J, Cornell Sarah E, Fetzer I, Bennett Elena M, Biggs R, Carpenter Stephen R, de Vries W, de Wit Cynthia A et al (2015) Planetary boundaries: guiding human development on a changing planet. Science 347:1259855. https://doi.org/10.1126/science.1259855

Teferi ZA, Newman P (2017) Slum regeneration and sustainability: applying the extended metabolism model and the SDGs. Sustainability 9:1–16

Teferi ZA, Newman P (2018) Slum upgrading: can the 1.5 °C carbon reduction work with SDGs in these settlements? Urban Plan 3:52–63. https://doi.org/10.17645/up.v3i2.1239

Thomson G, Newman P (2016) Geoengineering in the Anthropocene through regenerative urbanism. Geosciences 6:46. https://doi.org/10.3390/geosciences6040046

Thomson G, Newman P (2018) Urban fabrics and urban metabolism – from sustainable to regenerative cities. Resour Conserv Recycl 132:218–229. https://doi.org/10.1016/j.resconrec.2017.01.010

Thomson G, Newman P (2020) Cities and the Anthropocene: urban governance for the new era of regenerative cities. Urban Stud J Ltd 57:1502–1519. https://doi.org/10.1177/0042098018779769

Thomson G, Newton P, Newman P (2016) Urban regeneration and urban fabrics in Australian cities. J Urban Regen Renew 10:1–22

UNCRD (2019) Bridging the urban transport infrastructure gap in the context of smart and resilient cities – the role of private sector and public-private-partnerships. In: Proceedings of background paper for environmentally sustainable transport conference, Hanoi, Vietnam

Webber M (1963) Order in diversity: community without propinquity. In: Wingo L (ed) Cities and space. John Hopkins University Press, Baltimore

Webber M (1964) The urban place and the non-place urban realm. In: Explorations in urban structures. University of Pennsylvania Press, Philadelphia

Webber M (1968) The post city age. Daedulus 97:1093–1099

Wendling LA, Huovila A, zu Castell-Rüdenhausen M et al (2018) Benchmarking nature-based solution and smart city assessment schemes against the sustainable development goal indicator framework. Front Environ Sci 6. https://doi.org/10.3389/fenvs.2018.00069

West Australian Government (2019) Western Australian renewable hydrogen strategy and roadmap. Availabe online https://www.wa.gov.au/government/publications/western-australian-renewable-hydrogen-strategy-and-roadmap. Accessed 25 Mar 2022

Whitehead J, Newman P, Whitehead K (2022) Striking the right balance: understanding the strategic applications of hydrogen in transitioning to a net zero emissions economy. Sustain Earth. In Press

Wilson EO (1984) Biophilia. Harvard University Press, Cambridge, MA

Chapter 5
Smart Cultural and Inclusive Cities: How Smart City Can Help Urban Culture and Inclusion

Abstract Even though liveability was a major focus of ancient and modern urban thinking, with an aim of enhancing both societal and economic benefits, the consideration of cultural and inclusive attributes of cities is often overlooked. However, they are key in shaping people's sense of belonging and in building the identity of places; hence, they are an essential dimension in the pursuit of urban sustainability. The current concept of Smart Cities undermines those attributes and tends to portray and promote modernist principles that smart digital data alone is all that is needed for making good cities. Such ideas are counterproductive in the preservation and celebration of culture and can exclude many parts of society. This chapter explores the paradigm of culture and inclusion and builds the case that they are pivotal elements in achieving sustainability and how Smart City technologies can be used to enable their better integration.

5.1 Introduction

Good cities are those with a strong culture and are inclusive of all their residents and visitors; this is an ancient idea that has been necessary to rediscover over and over. In today's world, modernist ideals from the 1930s began to assert that new technology could enable good cities by clearing out the old cities of yesteryears and building new high-rise steel and concrete residences served by freeways that helped clean the slate from previous urbanism. The modernist city was driven by the need for mass production to create more housing for cities, especially those destroyed by two world wars. The modernist notions were motivated by equity and inclusion but did not have a strong link to culture, especially that which was expressed in the built form of older cities. The movement has been seriously criticised as outlined in previous chapters by Jacobs, Alexander and Salingaros, and a different approach based on respecting urban fabrics has been suggested as a better way to create a good city. However, modernism continues to drive much of the world's cities, especially in low-income urbanism where high-rise buildings have become a symbol of modern, wealthy futures.

Z. Allam, P. Newman, *Revising Smart Cities with Regenerative Design*, Cities and Nature, https://doi.org/10.1007/978-3-031-28028-3_5

The modernist city is not necessarily a Smart City as it is not about digital data; however, the modernist city became associated with the Smart City movement and raises issues about whether this association is helping or hindering in the creation of good cities. This chapter begins by addressing that issue. Then this chapter seeks to examine how culture and inclusion can be given higher priority in cities in ways that enable the regeneration of public spaces and the celebration of urban cultures in a fully inclusive way. This process is critical for the deeper values of cities that have been examined in the last three chapters and indeed are not just part of sustainability, net zero and regenerative design, but they are essential for being a good city. In this lens, we address culture within the built environment, i.e. narrowed to architecture and sites, responding to 'tangible' attributes.

The question of this chapter is *how Smart City technologies and systems enable culture and inclusion while also enabling sustainability, net zero and regenerative design outcomes*?

5.2 Modernism, Sustainability, Culture and Smart Cities

As the world witnesses an increase in urbanisation, there is a resulting boom in the construction industry aimed at trying to satisfy the increased demand for residential and commercial spaces. The UN Habitat (2020) notes that currently, the number of people living in urban areas is 55% and, by 2050, it will have risen to 70% (Kaneda et al. 2020; UN Habitat 2018; United Nations 2016), hence providing the construction industry with an increasing demand. City managers and developers respond to increasing demand by providing as much housing as possible on any available site, and this invariably has modernist construction techniques and architecture. These technologies have been beneficial to the construction industry where it is now possible to construct momentous structures with the potential to house hundreds of people and commercial activities in one building. The question is whether such construction is able to address issues of culture and sustainability and whether smart systems can help or not?

This modern approach of construction is hailed as the best way to proceed in low-income cities especially, but most fast-growing cities in high-income countries with spatial issues have also prioritised high-rise building (Alli and Moon 2018). The number of high-rise architectural buildings globally has increased exponentially. In 2017 alone, it is reported that approximately 144 skyscraper projects were completed in 69 cities distributed in 23 countries globally (Council on Tall Buildings and Urban Habitat (CTBUH) 2018). However, those can be detrimental to areas that are traditionally characterised as holders of rich historical heritage and where the fabric celebrates unique constructions of both historical and cultural value (Tangbin 2012). Even though this trend to modernism in architecture has often been criticised by many as being both unsustainable and unconducive to the preservation of culture (Al-Kodmany 2018; Pruetz 2017), there are others that believe that there is a cohabitation to be achieved (Rosen and Charney 2016; Appert and Montes 2015).

The primary argument against modernism is the pursuit of building cities characterised of high-rise building and other avant-garde structures ultimately leading to the creation of homogenous landscapes that overshadow the traditional heritage inherent to old and unique cultural fabric. For instance, an abundance of modern structures in London is deemed to be harming the historical landscape of the city (Milmo 2015). Similar challenges are reported in Istanbul, Turkey (Sev et al. 2011), Doha (Boussaa 2014) and Chicago (UNESCO 2017a) among many cities across the globe. Often critics attribute this pursuit to short-term planning policies that are synonymous with low-income and emerging economies who have little choice but to build such mass housing. But in high and emerging economies, the criticism is that modernist architecture is simply to portray themselves as having modern outlooks (Murray 2012). Mazzetto (2018) views the idea of replacing the historic cultural heritage of cities with modern architecture as trying to change local identity and culture, an approach that is very often contested (Boussaa 2017). A case in point is one advanced by Boussaa (2017) touching on the regeneration and revitalisation of Msheireb, Doha, Qatar. According to him, though this strategy is focused on giving the city a new look, the only way to discover the urban identity of this city is revisiting its roots and cultural practices. Ultimately, the consequences of such policies are to be thought from a long-term perspective. From this regard, it has been observed that there are substantial economic losses from the adoption of modernist principles in cities (Cheshmehzangi 2015). This argument is based on the fact that the cultural attributes from heritage buildings, and from old neighbourhoods, have unique attributes that are often sought from a business perspective, as they offer a lifestyle that is increasingly in demand (UNESCO 2016). Hegazy (2015) showcases that conservation efforts on historical buildings in most developing countries have had numerous positive benefits, namely, in their potential to attract more tourists, improve on the wellbeing of the locals by creating employment opportunities and creating more business opportunities such as in the hospitality industry and cottage industries that allow locals to manufacture local products in small industries.

The observation by Hegazy (2015) is of significance to developing economies, especially with the role of technology gaining an increased adoption in the cultural sphere. Just like some developing countries have managed to reap from their strategy of preserving the historical architectures, developing economies could also feather their economic fortunes by capitalising on their unique culture and identity. Instead of concentrating on erecting structures that identify them as modern like their developed counterparts, they can utilise the opportunities availed by advanced technologies to brand their cultural and historical heritage and position it as one of their economic strongholds. In particular, the adoption of the concept of Smart City would allow them to use technology in innovative ways and provide opportunities to raise funds to renovate some of the historical architectures and reuse them in areas like hospitality industry or museums like in the case of the Guggenheim Bilbao Museum.

5.2.1 What About Sustainability?

Chong and Balasingam (2018) emphasise that increasing the adaptive usability of such sites has a multiplier effect where other economic activities, based on local culture and talent, would emerge, hence, in the long run, promoting inclusivity and sustainability. De Cesari and Dimova (2019) are of the opinion that the adoption of technology can help spur the sustainable development and inclusive growth, especially if this is done to economically capitalise on the cultural heritage of the local neighbourhoods. To further encourage this, Pietro et al. (2018) highlight that the use of technology can help in financing mechanisms and further increase the democratisation in the way heritage resources are managed and allow for the trickling down effect of benefits to locals who are part and parcel of the heritage fabric. In Malaysia, for example, advanced technology has been integrated in the country's built cultural heritage, and this has allowed for the growth in the local communities, and as a result, the inhabitants are able to experience improved value in all aspects of the city fabric as reported by Bakri et al. (2015).

All these comments in the last paragraph are similar to the criticism levelled at Smart Cities in the last few chapters. To suggest that somehow modernism will magically provide sustainability outcomes simply by increasing economic activity or being more in tune with culture is just not right. There is no guarantee of this flow-on into sustainability. Technology needs to be given direction. These two comments are similar to the criticism levelled at Smart Cities in the last few chapters that somehow modernism will magically provide sustainability outcomes simply by increasing economic activity. Technology from smart systems, sustainability and culture need to be integrated into the design and procurement process based on the deeper values of a city. This chapter will seek to illustrate this further.

5.3 Culture and Sustainable Urban Development

The cultural identity intrinsic to cities differs as those are subject to different factors linked to political, economic and cultural dimensions. The United Nations (2015) emphasise that for cities to achieve sustainable development, these three factors need to be dwelled into but, most importantly, need to include the cultural factor, which unfortunately is not emphasised enough in urban regenerative frameworks and models (Lazaroiu and Roscia 2012). As is reported by the World Bank (2018), until some decades ago, cultural heritage was not considered as an important aspect in the development of cities. For this reason, the report documents that a substantial number of cultural heritage sites around the world had been neglected and some were destroyed in the favour of modern constructions in city upgrading programs since those were not considered to hold much significant value in cities. A similar observation is made in a report by Boussaa (2014) who argue many cultural sites are facing diverse challenges especially those that are geared toward adoption of modern urban trend.

Even though modernist approaches are negating the unique identity of cultural fabrics, the contribution of culture in urban areas is apparent (Allam and Newman 2018b; Labadi et al. 2021). It is now viewed as part and parcel of urban agendas, and as Wilson (2017) posits, culture is viewed as being a subset of the sustainable development goals (SDGs) and even eventually leads to their achievement. This proposition is clearly captured in a resolution made during the International Conference on Cultural Policies for Development (UNESCO 1998) in Stockholm back in 1998. The resolution proposed that policies on culture are to be considered as integral components of urban development agendas. This is also well captured in the 17 SDGs ought to be achieved by 2030. Though not explicitly mentioned as a stand-alone goal, as noted in a UCLG (2018) report, the concept of culture and its contribution in sustainability is captured in targets 4.7, 8.3, 8.9, and 11.4 of the SDGs. Its significances in cities are also captured in the New Urban Agenda (UN Habitat III 2016), where it is hailed as 'a source of enrichment for humankind and in the sustainable development of cities'. On the same, UNESCO has been in the forefront advocating for inclusion of cultural heritage in urban sustainable planning and policy discussion especially by noting how it contributes to the success of emerging cities and wellbeing of smaller urban vulnerable communities (UNESCO 2017b). UNESCO states that, by preserving and celebrating local cultures, economic activities powered by increased tourism, innovation and creativity and employment opportunities are bound to increase to the benefit of both governments and the local communities (Haans and Witteloostuijn 2016).

The inclusion of culture in urban development through projects and policies is thus paramount. This is so noting that the future of the world population is expected to reside in urban areas. Indeed, this number is foreseen to reach over 70% by 2050 and is understood to eventually lead to expand on issues like poverty, environmental degradation, scarcity of housing and the inequality in supply of basic resources. Nevertheless, UNCTAD (2018) and Oad et al. (2018), among others, note that by taping on the rich cultural diversity of cities, such problems can be overcome as this approach promotes creativity, job creation and diverse background exchanges, hence increasing innovation and creativity. By promoting the cultural dimension of cities, UNESCO (2009) found out that economic fronts like tourism, performing art and heritage and small manufacturing industries tied to culture and historical sites have the potential to contribute greatly to the growth of local and regional economies. This proposition is supported by the fact that these sectors were responsible for over 6.1% growth in the global economy, which is equivalent to $ 4.3 trillion per annum globally, in 2015, a percentage that has risen to over 7.9% as of 2017 of the global economy and job opportunities for over 30 million people. In addition, the cultural industry has the potential to foster social cohesion and integration, help in strengthening of cultural diversity and improve on national identity and attract foreign direct investments (FDIs), among many other benefits (Fonseca et al. 2018; Vecco and Imperiale 2018; NatalinaCarrà 2016; Caust and Vecco 2017).

The above benefits from the inclusion of culture in urban planning are not overstated, as those have been tried and tested in numerous cities, and have been known to improve on liveability levels, relative to modern cities. A notable example is Central Norway, which is home to Røros, a World Heritage Site characterised by dense and historic wooden urban structures (Bye 2008). It is argued that the presence of this site has promoted a compact urban centre with locals reporting high levels of satisfaction. Another example is Mexico City, which prides a myriad of historical buildings and a rich cultural heritage that has been outstanding in promoting its economic agendas (Garcia and Piccio 2018). Similarly, it is said that the cultural landmarks, like the historic buildings that uniquely identify Mexico City, have served as a social bond. Other cities that are within this category include Barcelona, London, Amsterdam (Angelidou et al. 2017b), Porto (Liberato et al. 2018), Tokyo and a score of Chinese cities (Yang and Černevičiūtė 2017), among many others. However, there are numerous cities where the cultural dimension is not capitalised on and slowly facing a state of decay, thus losing on economic benefits along with the identity of place. An example of this is the small island nation of Mauritius (Allam 2018).

Though some quarters have advanced that culturally oriented cities have higher developmental potential as opposed to modern ones, it is believed that the cohabitating of best practices from the two school of thoughts can reap higher benefits (Siew and Allam 2017). This is true especially by leveraging on the power of technology to bring the best of cultural attributes and integrate it with modern trends like the emerging concept of Smart City. Angelidou et al. (2017b) posit that such a model that marries the two distinct lines of thinking can benefit urban areas by bringing prosperity, efficiency and competitiveness in different fronts while, at the same time, maintaining a unique identity highlighted by local culture and heritage. The city would then benefit by having its cultural heritage preserved and buildings renovated and reused to maximise on their potential and, at the same time, pacing toward sustainable development and improving on economic growth (Franco 2018).

5.4 Cultural Heritage and the Economy

The contribution of cultural heritage in society is generally considered to be immeasurable and insatiable. EY (2015) argues that, together with creativity, cultural heritage serves as the societal binding glue especially noting that it has a relationship with politics, society, environment, economy and technology. The ability to link the past, the present and the future furthermore makes culture a strong societal component that needs to be placed at the centre stage of any policy framework.

The investment in cultural heritage is seen as one of the most potent strategies that a city can employ to promote its job creation strategies, increase income flow and foster competitiveness of the urban fabric to support local policies of economic empowerment and resilience (Lavarone et al. 2019; Allam and Newman 2018a, b; Allam 2012). Bowitz and Ibenholt (2006) argue that by promoting cultural heritage and conserving historical sites, activities celebrating the urban fabric such as local festivals, amusement parks and culturally oriented products can be made to mushroom, hence increasing tourism activities both from domestic and foreign visitors and impacting on the local and regional economy.

Vegheş (2018) suggests that the advancement in technology has allowed different cities and governments to market their cultural heritage, thus evoking interest among people, irrespective of locality and context to learn more about, and to seek, cultural experience. This is accentuated by a flourishing economy surrounding the urban cultural dimension. An OECD (2018b) report documented that the marketing strategy on cultural goods and services is enabled by modern trends like digitalisation and globalisation. Those enable access to technologies applicable to the cultural sphere, at almost real time and at a reduced cost. The report advances that these trends are made possible by the fact that culture is now already well integrated in areas like education, health and welfare and economic sectors where it plays an instrumental role, especially in improving quality of life. Lekaota (2018) adds that, unlike in the past where only locals or central governments were benefiting from most cultural heritage products and services, the increased profit-sharing strategies adopted in different places incorporating even the local communities are yielding positive results which are benefiting immediate economies. Given this, there are numerous creative and innovative products and services currently being developed around the world, which are helping to fuel conservation efforts and are widely praised. To affirm this, a UNESCO report (Intellectual Property Watch 2016) highlights that, in the recent past, the number of cultural products and services being patented is increasing and ranges from artefacts, songs and music, festivals and games, documentaries, jewelleries and other wearable and historical sites, among many others. This highlights the increasing interest, adoption and economic potential of such approach.

EY (2015) further posits that cultural heritage does not only help local communities to improve on liveability but is now being also used as a broader tool for urban regeneration. For instance, the city of Bilbao, Spain, stands as a testament to this following the construction of the Guggenheim Museum. Reports (Plöger 2008; EY 2015) indicate that this city was marred by many challenges like petty crime, inadequate living conditions, uncontrolled immigrant unemployment, crisis and unemployment with only 25% of its habitats employed in the 1990s. Nevertheless, though deemed as an expensive expedition, the revitalisation of Guggenheim Museum significantly addressed unemployment through a number of factors. First, after completion, the project employed over 1000 full-time workers. Second, the museum attracted numerous other economic activities like in the hospitality industry due to its tourist attractiveness which have high traffics since then as depicted in the Fig. 5.1. In fact, even after the impact of COVID-19, the museum is reported to have started recovering as from 2021, and it is projected to continue attracting more visitors post-pandemic. Plaza and Haarich (2013) highlight that such increase is prompted by the branding effect of the museum, placing it in the league of global touristic destination of choice.

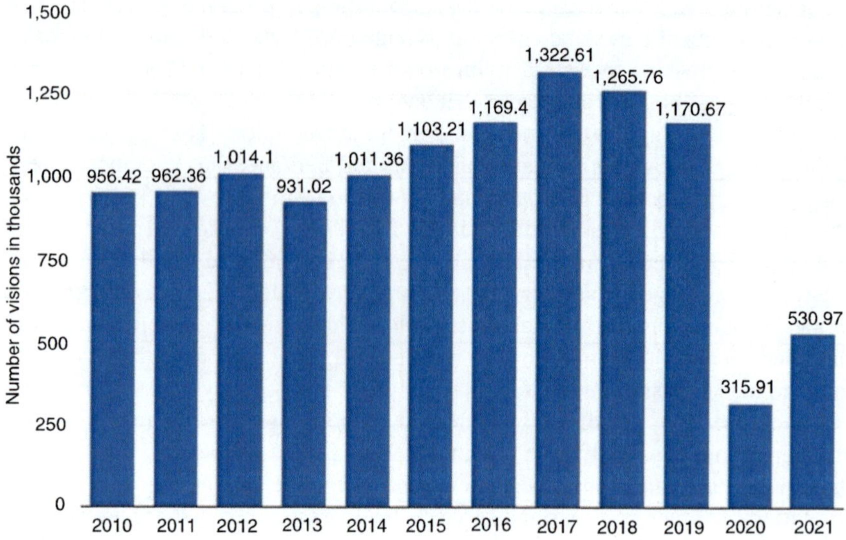

Fig. 5.1 Annual visitors to the Guggenheim Museum in Bilbao, Spain, from 2010 to 2021. (Source: Statista 2019; López 2022)

The overall impact of the revitalisation of the museum is the regeneration of the city of Bilbao from its previous status to a modern city that has the potential to support its residents and visitors within its fabric. Kaczmarska and Kaczmarska (2018) note that the museum humanised the city, fostered its attractiveness and created opportunities for modern solutions to be integrated in architectural and design works targeting different aspects of the urban fabric.

Another city capitalised on its cultural heritage is the city of Nantes. Having its historical identities, like its shipbuilding industry collapse, the city was under severe economic stress (Nicolas 2014). Nevertheless, its fortune changed when 'La Machine', a street theatre company, decided to revive the historical shipbuilding yards and warehouses with a different economic activity while capitalising on the cultural strength of the locality along with innovation and creativity (Dunmall 2016). This started with a festival which incorporated theatre, art and music and later grown to internationally attractive events like the 'Voyage à Nantes' (Mansfield 2017), which is said to have attracted economic returns of over €48.8 million in 2016. Interestingly, through the cultural sphere, the city of Nantes has embraced modern trends like Smart City infrastructures coupled with world-class hospitality infrastructures, and the city is now taunted as among the fastest growing in France due to its liveability level.

As noted by UNESCO (2016), various cities around the world have adopted culture as a pivotal pillar of their economy. This fact is supported by increasing activities linked to the improvement of cultural heritage and historical sites, and Soutinho (2018) further shares that the contribution of culture in improving the urban values and the conservation and preservation of cultural heritage needs to be

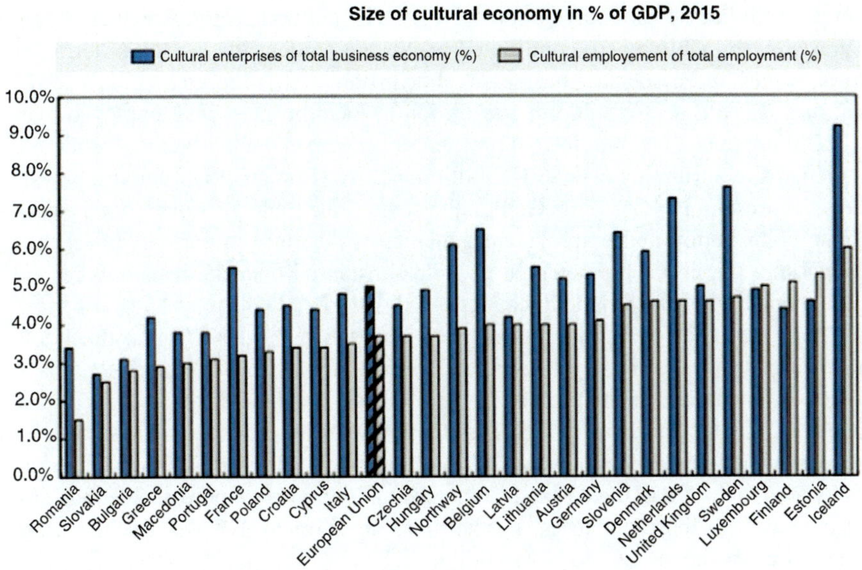

Fig. 5.2 Size of the culture economy in the European Union. (Source: OECD 2018a)

at the forefront of urban regeneration conversations. He argues that through culture, small and medium-size cities can join their larger ones in maximising their cultural endowment to compensate for resource limitations. The main attraction in culture is its ability to spur increased job opportunity, promote local business and promote cultural tourism, which is benefiting many cities. For instance, the reconstruction and rehabilitation of such historic sites like the Sankore Mosque in Timbuktu, Mali, which was destroyed in 2012 (Jarus 2013), and the Al-Askari Shrine, Samarra, Iraq, have attracted not only financial benefits but contributed to restoration of social cohesion, a factor that is paramount in the economic growth of a city.

Interestingly, it is seen that countries benefit greatly from cultural economy. Figure 5.2 showcases the contribution of culture to the various countries of the European Union.

5.4.1 *Can Culture and Heritage Prompt Enough Capital for Smart City Concept Implementation?*

Soutinho (2018) outlines a list of cities like the city of Baena, Spain; Dodoni, Greece; Cahors, Southern France; Espinho, Portugal; Montova, Italy; and Armagh, Northern Ireland which are capitalising on their cultural heritage to improve their economy. California City, which is home to Hollywood, is also a classical example of how much culture can contribute to the economy. It is reported that in 2018, its gross domestic product (GDP) was over $2.7 trillion, a figure that makes it the world's fifth largest economy (Cooper 2018). The cultural sector, which is largely represented by Hollywood, contributed a sizeable amount of revenue in terms of taxes, wages and associated businesses. For instance, it is reported (Busch 2018) that in 2018, this sector paid out $49 billion to local business and employed over 2.1 million people. Overall, the cultural sector is deemed one of the most profitable in the USA, contributing over $804.2 billion in 2016 (Bureau of Economic Analysis 2019), representing 4.3% of its GDP in the same year. Table 5.1 represents the economic contribution and employment generated by culture in various cities, thus showcasing that it has the potential to yield substantial financial resources for the implementation of the Smart City agendas.

There are other cities distributed in countries like the UK, China, the USA, Australia, Malaysia, Turkey, Egypt and India, among many others, that have advanced and branded themselves by capitalising on their creative and cultural industries (CCI) (Garcia and Piccio) as shared by Oad et al. (2018). These authors report that CCI contribute to over 4.2%, 5.2% and 6.37% of the GDP in diverse cities located in the USA, the UK and China and have employed more people than any other sectors in these countries. Furthermore, a report by EY [25] evaluates the global contribution of CCI to $2250 billion.

Table 5.1 Economic contribution and employment generated by culture

City	Economic contribution from cultural heritage	Employment from cultural heritage	Sources
California	$49 billion	2.1 million	Cooper (2018)
St. Louis	$590.9 million	19,129	Hellmuth (2017)
Rome	Not available	250,300	Antorio Arts Council (2018)
Toronto	$11.3 billion	174,000	Toronto (2017)
Ontario	$25.7 billion	269,000	Antorio Arts Council (2018)
London	£3.7 billion	196,000	World City Culture Forum (2018) and Cebr (2018)
Brussels	Not available	76,400	World City Culture Forum (2018)
Bueno Aires	Not available	28,760	World City Culture Forum (2018)
Cape Town	Not available	83,490	World City Culture Forum (2018)
Dublin	Not available	6000	World City Culture Forum (2018)
Helsinki	Not available	32,170	World City Culture Forum (2018)
Hong Kong	5% or $17.02 billion	40,400	World City Culture Forum (2018) and Hong Kong Economic (2018)
Los Angeles	$115 billion	1.6 million	World City Culture Forum (2018) and Otis Collage of Art an Design (2018)
Melbourne	$23 billion	202,760	World City Culture Forum (2018) and Cuthbertson (2017)
Zürich	Not available	38,188	World City Culture Forum (2018)

5.5 Cultural Heritage and Technology

The marriage of culture and technology has allowed for unbarred innovation and creativity which in turn has positioned cultural products and services as economic pillars and has opened opportunities to expand those beyond physical items. The EY (2015) report denotes that through technology, the cultural world is now becoming more diversified and competitive. For instance, prior to 2015, physical musical products had been dominant, but this changed in favour of the digital music market (Hassan 2016). It is further recorded that in 2015, the digital cultural segment accounted for over 45% of all the sales, while physical goods trailed only at 39%. In 2018, Hernandez (2018) further reported that the digital market had risen to 87% (75% streaming and 12% download), with the physical market only accounting for a low of 10% of the total music sale. The same trends are real in the sale of other products like books, videos and games, among many others. The find of this report attributes these successes in digital cultural products to ease of entry in the market enabled by unlimited potential of technology and its advancement thereof. One such potential is the idea of digital distribution where event organisers are reported to use cultural data to promote events and also to sell the e-tickets to targeted groups and consumers.

In addition, technology has allowed for the creation of progressive products and services in various realms such as virtual reality, 3D modelling, augmented reality and others (Ramos et al. 2018; Jung and Dieck 2017). Through these platforms, products such as advanced video games and state-of-the-art mobile applications have been developed and spurred cultural heritage likeability to new heights. For instance, in Italy, a country with rich culture, especially captured well in some of its iconic historical sites like the Coliseum, there is a new technologically powered initiative by the name of *ARmedia 3D Tracker* that allows visitors to visually, in a 3D representation, experience the Coliseum by reconstructing the missing pieces that were part of this iconic structure many years back (JCDecaux 2014). Through this platform, visitors are able to visualise how the entire building was rendered during its glorious days (Zarzycki 2016). A similar application is the *Imayana*, which allows a virtual tour of Bordeaux which was in operation during the eighteenth century (Cloutet 2012). Google Street View is another wonderful application that has been used to give its users virtual image of how cities and allows for an increased audience to experience the unique fabrics of old towns and cities (Dredge 2016).

Going further from the heritage component, it is also possible, through Google Street View, to have a virtual tour of how the Freedom Tower in New York looked at every stage of its construction till its current form. Those tools are however not used only as recreational elements and are even used for educational purposes. Mortara et al. (2014) share how *serious games* (games meant for educational purposes) are now being used as learning tool for different cultural contents, hence exposing various cultures in both an entertaining and educational manner. In light to this, statistics shows that there are nearly 3 billion video gamers globally, rendering consumer spending to nearly $78 billion, and this figure is projected to rise abruptly in coming years (Ivan 2022). Therefore, capitalising on this surge and from the interest of the gaming industry into historic screenplays (e.g. Assassin's Creed), industries linked to cultural heritage is bound to grow as more types of digital products are introduced.

Cities, especially those that have already embraced the concept of Smart Cities, have the potential to utilise technology to further integrate cultural dimensions, and those could be further accentuated by the adoption of a series of fiscal incentives as in the case of Mauritius (Allam and Newman 2018a; Allam et al. 2018). These technologies could allow for cultural data gathering and sharing via the different platforms and, by so doing, help in promoting economic activities linked with culture. If planned correctly, it can thus be seen that the construction of smart infrastructures with optimal use of resources and the improvement in the security sectors, among others, can be made to support cultural dimensions of cities. For instance, the technologies are paramount in the rehabilitation and preservation efforts and also in securing historical sites at risk from terrorists, illicit traffickers of cultural products like artefacts or weather-related incidences like flooding and earthquakes. Soutinho (2018) asserts that these technologies have the potential to advance agendas and conversations that are geared toward encouraging maximum utilisation of cultural resource and, in the course of doing this, promote revitalisation and reconstruction efforts.

5.6 Cultural Heritage and the Smart City

Even though the role of technology is clearly beneficial for the promotion of culture, it is however not always implemented in technologically driven initiatives. In support of this proposition, Franco (2018) argues that, in more cases than not, especially in cities, the focus of managements while introducing new technologies in their jurisdiction is geared toward infrastructure, security, housing and combating impacts of climate change, among other challenges, but rarely is directly linked to cultural heritage. He suggests that most technologically oriented initiatives, unless adequately addressed, are not compatible with cultural attributes, especially in regard to restoration and rehabilitation.

Some scholars (Klimovský et al. 2016; Azad et al. 2010) believe that cultural practices and heritage can be interpreted as being obstacles and as a resistance to the adoption of technology and to its advancement. For instance, Cameron and Kenderdine (2010) argue that technology is sometimes perceived in some quarter as counterintuitive, hence, limiting access to material culture, and others believe that technology has the potential to raise concerns over 'ownership of heritage'. Klimovský et al. (2016) argue that, in instances where the public is not well versed or informed about the role of technology, especially with issues that have to do with their culture, they are bound to resist it by citing concerns related to security, control and unwarranted use of their data. However, Lewis et al. (2018) suggest that cultural heritage, through its diverse unique characteristics, has the potential to foster increased social inclusion and promote people's identities and, in the course of doing this, allow for integration of technologies. In particular, in the recent past, with increasing loss of heritage linked to terrorism activities in cities, there is increased application of technology to secure and preserve architectural heritage that is at great risk of being lost completely. For instance, it has been found that through technologies such as cloud services, it is now possible to store and share information and details about how different cultural heritages have been transformed, an exercise that was inclusively confined in the hands of archives, hence rendering it inaccessible to the public (Riganti 2017; Allam and Dhunny 2019).

Angelidou et al. (2017a) explain that a majority of local authorities and governments opt for Smart City concepts and rely on technology, to achieve prosperity, effectiveness, sustainability and competitiveness in all aspects of their cities. Though similar objectives are advanced in the pursuit of cultural heritage management, it has been observed that a majority of Smart City projects barely integrate cultural dimensions in their endeavours (Fernandez-Anez et al. 2018b; Riganti 2017). This is so despite the fact that the term 'cultural heritage' encompasses diverse categories of heritage. A UNESCO (2017b) report outlines these categories in the form of (i) tangible (movable, immovable, underwater) and (ii) intangible (UN Habitat III) cultural heritage. In cities, one or more of these categories of cultural heritage may be present and need to be considered during the planning and implementation stage of Smart City technology implementation so as to optimise the experience of the residents. Nevertheless, a survey of literature showcases that despite the importance of those cultural elements, the cultural theme is mostly *not* in the numerous Smart City models as shown in Table 5.2.

Table 5.2 Inclusion of culture as a primary dimension in Smart City models

	Inclusion of culture	
Smart city model	Yes	No
Lazaroiu and Roscia (2012)		X
Sepasgozar et al. (2019)	X	
Fernandez-Anez et al. (2018a)		X
Psomadaki et al. (2018)		X
Andrisano et al. (2018)		X
Ferrer (2017)		X
Cugurullo (2017)		X
Yigitcanlar et al. (2018)		X
Yigitcanlar and Lee (2014)		X
Zvolska et al. (2018)		X
Caprotti (2018)	X	
De Azambuja et al. (2014)		X
Cowley and Caprotti (2018)		X
Angelidou et al. (2017a)	X	
Caird and Hallett (2018)	X	

Despite the push and pull on the need to include culture in the Smart City implementation agenda, it has been found that its inclusion has far more benefits in those cities that are characterised by rich cultural fabrics. Vecco and Imperiale (2018) suggest that cultural heritage is a nonrenewable resource that cannot be replaced once destroyed, and this quality positions them as key contributors to economic growth. This fact is tied to their unmetered quality in attracting tourism activities, creating job opportunities and fostering social cohesion. However, these authors acknowledge that cultural tourism is threatened by factors like wars, illegal trafficking of cultural products, uncontrolled tourism activities and limited budgets. In the recent past, due to the increased attack on cultural heritage, multifaceted efforts to reverse the said factors have been initiated in different quarters (Bokova 2017; Weiss and Connelly 2019), and this is bound to benefit cities that have factored in the cultural aspects in their plans. Psomadaki et al. (2018) explain that such benefits arise partly in the form of increased funding from international originations such as UNESCO in their bid to help prevent destruction and also help economies to attract foreign direct investments through, and in, the cultural heritage dimension of cities.

Angelidou et al. (2017a) are of the opinion that Smart Cities that have placed emphasis on cultural heritage benefit from the citizen acceptance of their project. This is informed by the fact that most Smart City projects are geared toward addressing local needs and challenges and development demands that are prompted by the cultural identity of cities (Allam and Dhunny 2019; Allam 2018). Marsella and Marzoli (2017) argue that among the local demands is the preservation of local culture, historical sites and cultural products, and with the Smart City technologies such as big data and artificial intelligence, they argue that such demand would be easily met. The authors explain that where cultural dimensions are already part of the Smart City agenda, there would be reduction of costs, especially those dedicated in conservation and rehabilitation of historical sites and in digitisation of different cultural aspects. Thus, integrating culture in Smart City agenda is also taunted as a smart move in addressing the challenge of housing and scarcity of space (Allam and Jones 2019). In particular, this is possible through the reconstruction, rehabilitation and reuse of some of historical site for other activities with higher economic returns, but those that are not bound to compromise the cultural aspect of the infrastructure (Farrag and Abouhadid 2018).

Culture can thus be shown to support the increase of liveability levels and could equally be combined to Smart City models which are also portrayed to support the same goals. A combination of both could potentially result in better outcomes to support increasing sustainability, liveability and inclusivity levels.

Culture and creativity are closely linked, and in Newman (2020), the idea of a new kind of creativity emerging as part of the new regenerative economy is outlined. The notion of a creative city is developed where indigenous place-based urbanism is integrated into the way that nature-based planning is able to create the more biophilic and wild aspects to a city as well as something deeply beneficial to community growth.

5.7 Slum Regeneration, Community and Smart City

In 2003, there were an estimated 1 billion slum dwellers in the world; by 2020, this figure doubled to two billion people (UN Habitat 2022). Urbanisation is a global phenomenon, sometimes leading to the dramatic growth of cities and urban districts which can stretch the ability of the city to provide sufficient adequate housing. In developing countries, informal urbanism emerges from their massive rural-urban migration movements. Currently, an estimated 72% of urban dwellers in sub-Saharan Africa are living in such informal slums. With only an estimated 24% of the population living in urban areas, East Africa remains the least urbanised subregion in Africa, so if that changes, the amount of informal settlement in Africa may grow even faster.

Slum redevelopments have had mixed success in improving the sustainability of human settlements. Slum settlements in many emerging cities face many challenges due to their economic status and environmental problems; however, they have a major benefit: their culture of social cohesion is very high. So is it possible that this foundational characteristic can be better accommodated in any slum settlement schemes?

Teferi and Newman (2017) have used the extended metabolism model (see Fig. 5.3) to assess the resource consumption, wastes and liveability outcomes in a study that compares the informal slum settlements in Addis Ababa with high-rise slum clearance apartments. This kind of modernist high-rise approach has been used as the main slum clearance model for the world's informal settlements which are called the organic model in the study by Teferi and Newman (2017).

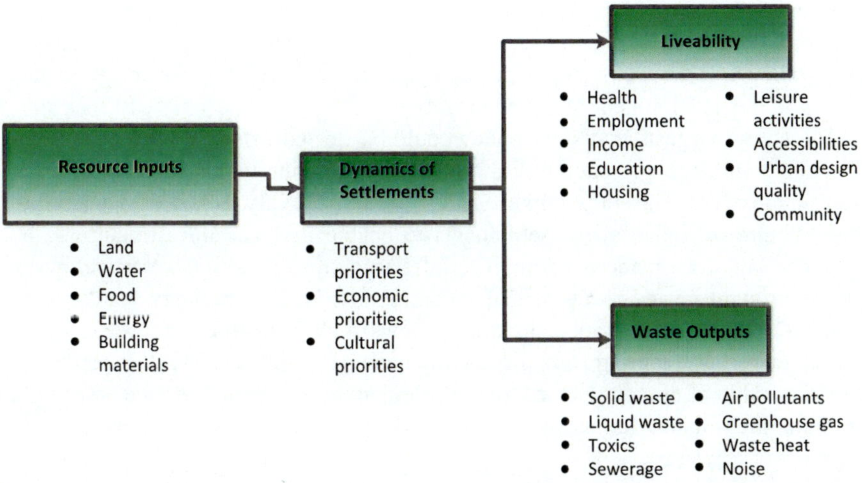

Fig. 5.3 Extended metabolism of Addis Ababa organic slum settlements of modernist high-rise slum clearance settlements

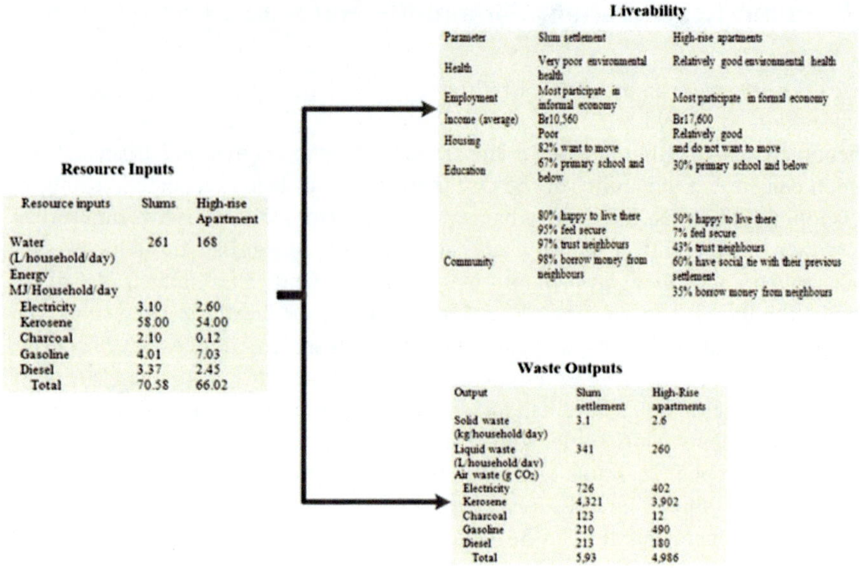

Resource Inputs

Resource inputs	Slums	High-rise Apartment
Water (L/household/day)	261	168
Energy MJ/Household/day		
Electricity	3.10	2.60
Kerosene	58.00	54.00
Charcoal	2.10	0.12
Gasoline	4.01	7.03
Diesel	3.37	2.45
Total	70.58	66.02

Liveability

Parameter	Slum settlement	High-rise apartments
Health	Very poor environmental health	Relatively good environmental health
Employment	Most participate in informal economy	Most participate in formal economy
Income (average)	Br10,560	Br17,600
Housing	Poor 82% want to move	Relatively good and do not want to move
Education	67% primary school and below	30% primary school and below
Community	80% happy to live there 95% feel secure 97% trust neighbours 98% borrow money from neighbours	50% happy to live there 7% feel secure 43% trust neighbours 60% have social tie with their previous settlement 35% borrow money from neighbours

Waste Outputs

Output	Slum settlement	High-Rise apartments
Solid waste (kg/household/day)	3.1	2.6
Liquid waste (L/household/day)	341	260
Air waste (g CO_2)		
Electricity	726	402
Kerosene	4,321	3,902
Charcoal	123	12
Gasoline	210	490
Diesel	213	180
Total	5,93	4,986

The results show very little difference in resource consumption and waste produced but show liveability outcomes are mixed: economic benefit is substantially improved in the high-rise areas due to becoming part of the formal economy, but community networks and trust are substantially lost when people transfer from the slums. Thus, all the community benefits that are in reality the culture of the organic settlement are lost when the community is broken up.

The economic outcomes from the new high-rise development are significantly better than the informal slums, probably because the residents are given a formal address and hence can access the formal economy in terms of banking, education and employment. It is critical that this element is maintained in any slum management policy as it is a major step in providing people to end extreme poverty and also to help in the development of the whole city.

The study suggests that slum policy could be shifted from the modernist high-rise slum clearance approach to a more organic community-based renewal of the slums themselves. The study suggests that indeed Smart City technology can fit into the structure of the informal settlements to enable its basic infrastructure to be upgraded and even made net zero. The infrastructure for energy, water and waste can be brought in and made part of a community. New technology that fits into community-based governance structures allows such infrastructure to be a viable option as well as enabling formal economy benefits as the small-scale systems are just as capable of enabling an informal settlement to be formalised (see case study on slum in Malang, Indonesia, using small-scale sewage works; Newman and Kenworthy 1999a, b).

Smart city systems can be used to assist the management of energy, water and waste in small-scale distributed systems. Metering can be linked to a mobile phone and enable a system for managing the local area while also having government oversight through a digital link. The emergence of such demonstrations is something that can be linked to climate change funds or other aid projects.

Slums play an important role in building the city economy, in particular through the hard work of new migrants and their work in the informal sector, which is a vibrant support to the city economic system. The total contribution to employment by the formal sector in Addis Ababa is only 30%, the balance coming from the informal sector which is said to account for 70% of total employment. Thus, any disruption in the informal sector has implications for the city and its functioning. By providing a formal address and upgrading process in either the organic in situ approach or the modernist high-rise approach, the economic outcomes are likely to be similar.

However, the biggest difference between the slums and the high-rise projects is the astonishing difference in community trust and networks between the two settlement types. In the slums, people feel secure and borrow tools and money from the neighbours, but this is much less in the high-rise projects. There are ways this can be understood in terms of design where simple, walkable links are available between slum houses, but in high rise, the units are separated with little consideration of the need for community. There are architects who do find ways of building more community into high-rise developments, but as outlined, it may be a better approach to create organic in situ development that builds on the social capital in communities while enabling better metabolism and better economic outcomes.

The organic model based on in situ slum upgrading is an instrument that promises to promote empowerment, integrated urban development and social cohesion as well as the environmental upgrading so eagerly desired and the opportunity to enable economic development. This is a more compelling approach against the backdrop of ever-increasing urban unemployment, poverty and widening socioeconomic inequalities as well as the lack of government funds for large-scale interventions. To do so will need some appropriate Smart City technology to enable management of the small-scale systems, like rooftop energy production.

In order to achieve sustainable slum improvement, planners and city policy makers should recognise the value in slum settlement patterns and seek to build on the social capital that exists inside the urban fabric. In situ community-based, comprehensive slum upgrading does just that. There are now technologies for infrastructure upgrading using Smart City and small-scale net zero energy, water and waste systems that can make it possible to maintain the community and unleash a process of upgrading that is self-sustaining rather than requiring massive government intervention as seen in the slum clearance programs.

5.8 Conclusion

This chapter examined the role of culture as a driver for sustainable urban development and underlined how it has contributions impacting on all dimensions of the economy, society and environment, which can lead to the promotion of liveability within cities through social inclusiveness and citizen participation. This has been outlined in low- and high-income cities. Smart City technologies are now critical to enabling many of these desired outcomes. While those outcomes are seen as promising dimensions, it is unfortunately not well adopted in the contemporary notion of Smart Cities. The technology inherent to Smart Cities can support cultural dimensions and can thus further enhance its contribution to urban fabrics in terms of job creation, revenue generation and sustainability outcomes. Smart City redefinition needs to include the cultural and inclusive community values of regenerative urbanism.

References

Al-Kodmany K (2018) Sustainability and the 21st century vertical city: a review of design approaches of tall buildings. Buildings 8(8):102–142. https://doi.org/10.3390/buildings8080102

Allam Z (2012) Sustainable architecture: utopia or feasible reality? J Biourban 2:47–61

Allam MZ (2018) Redefining the smart city: culture, metabolism and governance. Case study of Port Louis, Mauritius. Thesis, Curtin University, Perth

Allam Z, Dhunny ZA (2019) On big data, artificial intelligence and smart cities. Cities 89:80–91. https://doi.org/10.1016/j.cities.2019.01.032

Allam Z, Jones DS (2019) The potential of blockchain within air rights development as a prevention measure against urban sprawl. Urban Sci 3(1):38

Allam Z, Newman P (2018a) Economically incentivising smart urban regeneration. Case study of Port Louis, Mauritius. Smart Cities 1(1):53–74

Allam Z, Newman P (2018b) Redefining the smart city: culture, metabolism & governance. Smart Cities 1:4–25

Allam Z, Dhunny A, Siew G, Jones D (2018) Towards smart urban regeneration: findings of an urban footprint survey in Port Louis, Mauritius. Smart Cities 1(1):121–133

Alli MM, Moon KS (2018) Advances in structural systems for tall buildings: emerging developments for contemporary urban giants. Buildings 8(8):104. https://doi.org/10.3390/buildings8080104

Andrisano A, Bartolini I, Bellavista P, Boeri A, Bononi L, Borghetti A, Vigo D et al (2018) The need of mutlidisciplinary approaches and engineering tools for the development and implementation of the smart city paradigm. Proc IEEE 104(4):738–760. https://doi.org/10.1109/JPROC.2018.2812836

Angelidou M, Karachaliou E, Angelidou T, Stylianidis E (2017a) Cultural heritage in smart city environments. Paper presented at the 26th international CIPA symposium, Ottawa, Canada, 28th August – 1st September 2017

Angelidou M, Karachaliou E, Angelidou T, Stylianidis E (2017b) Cultural heritage in smart city environments. Paper presented at the the international archives of the photogrammetry, remote sensing and spatial information sciences, Ottawa, Canada, 28th August – 1st September

Antorio Arts Council (2018) Economic contribution of arts and culture in Ontario. http://www.arts.on.ca/research-impact/research-publications/economic-contribution-of-arts-and-culture-in-ontar. Accessed 25 Mar 2019

Appert M, Montes C (2015) Skyscrapers and the redrawing of the London skyline: a case of territorialisation through landscape control. Artic J Urban Res Online. https://doi.org/10.4000/articulo.2784

Azad B, Faraj S, Goh JM, Feghali T (2010) What shapes global diffusion of e-government: comparing the influence of national governance institutions. J Glob Inf Manag 18(2):85–104

Bakri AF, Ibrahim N, Ahmad SS, Zaman NQ (2015) Valuing built cultural heritage in a Malaysian urban context. Soc Behav Sci 170:381–389. https://doi.org/10.1016/j.sbspro.2015.01.048

Bokova I (2017) UNESCO's role in emergency situations: what difference can soft power make in times of crisis. J Int Aff 70(2):59–68

Boussaa D (2014) Al Asmakh historic district in Doha, Qatar: from an urban slum to living heritage. J Archit Conserv 20(1):2–15. https://doi.org/10.1080/13556207.2014.888815

Boussaa D (2017) Urban regeneration and the search for identity in historic cities. Sustainability 10:48. https://doi.org/10.3390/su10010048

Bowitz E, Ibenholt K (2006) Economic impacts of cultural heritage – research and perspectives, Working Paper. Econ Analysis, Oslo

Bureau of Economic Analysis (2019) Arts and cultural production satellite account, U. S. and States 2016. BEA: U.S. Department of Commerce. https://www.bea.gov/news/2019/arts-and-cultural-production-satellite-account-us-and-states-2016. Accessed 18 Mar 2019

Busch A (2018) MPAA: hollywood is key driver of U. S. economy with $49 bn payout, salaries much higer thatn national average. Deadline. https://deadline.com/2018/01/film-and-tv-industry-49-billion-payout-u-s-economy-average-salary-42-percent-higher-than-national-average-1202244209/. Accessed 17 Mar 2019

Bye M (2008) A changing townscape: Røros throughout a century of heritage management. HiST – University College of Southern Trøndelag, Trondheim

Caird SP, Hallett SH (2018) Towards evaluation design for smart city development. J Urban Des:1–24. https://doi.org/10.1080/13574809.2018.1469402

Cameron F, Kenderdine S (2010) Theorizing digital cultural heritage. MIT Press, Cambridge

Caprotti F (2018) Spaces of visibility in the smart city: flagship urban spaces and the smart urban imaginary. Urban Stud. https://doi.org/10.1177/2F0042098018798597

Caust J, Vecco M (2017) Is UNESCO World Heritage recognition a blessing or burden? Evidence from developing Asian Countries. J Cult Herit 27:1–9. https://doi.org/10.1016/j.culher.2017.02.004

Cebr (2018) The heritage sector in England and its impact on the economy: a report for historic England. Center for Economics and Business Research, London

Cheshmehzangi A (2015) Urban identity as a global phenomenon: hybridity and contextualization of urban identities in the social environment. J Hum Behav Soc Environ 25(5):391–406. https://doi.org/10.1080/10911359.2014.966222

Chong KY, Balasingam AS (2018) Tourism sustainability: economic benefits and strategies for preservation and conservation of heritage sites in Southern Asia. Tour Rev. https://doi.org/10.1108/TR-11-2017-0182

Cloutet P (2012) Imayana Bordeaux discovery in augmented reality – Imayana Bordeaux XVIII. http://www.aquitaineonline.com/actualites-en-aquitaine/gironde/imayana-bordeaux/imayana-bordeaux-xviiieme.html. Accessed 17 Mar 2019

Cooper JJ (2018) California now world's 5th largest economy, surpassing UK. USA Today. https://www.usatoday.com/story/news/nation-now/2018/05/05/california-now-worlds-5th-largest-economy-beating-out-uk/583508002/. Accessed 17 Mar 2019

Council on Tall Buildings and Urban Habitat (CTBUH) (2018) 2017: skyscraper history's tallest, highest – volume, and most geographically diverse year. CTBUH J 1(1):44–51

Cowley R, Caprotti F (2018) Smart city as anti-planning in the UK. Environ Plan D Econ Space. https://doi.org/10.1177/2F0263775818787506

Cugurullo F (2017) Exposing smart cities and eco-cities: Frankenstein urbanism and the sustainability challenges of the experimental city. Environ Plan A Econ Space 50(1):73–92

Cuthbertson D (2017) Cultural tourism and creative industries pump $23b into Victorian economy. The Sydney Moring Herald. https://www.smh.com.au/entertainment/art-and-design/cultural-tourism-and-creative-industries-pump-23b-into-victorian-economy-20170620-gwuk7s.html. Accessed 25 Mar 2019

De Azambuja LS, Lheureux-De-Freitas J, Moreira CR, Macadar MA (2014) A smart city initiative. In: Proceedings of the 15th annual international conference on digital government research 14. https://doi.org/10.1145/2612733.2612768

De Cesari C, Dimova R (2019) Heritage, gentrification, participation: remaking urban landscapes in the name of culture and historic preservation. Int J Herit Stud 25(9):863–869

Dredge S (2016) The complete guide to virtual reality – everything you need to get started. The Guardian. https://www.theguardian.com/technology/2016/nov/10/virtual-reality-guide-headsets-apps-games-vr. Accessed 20 Mar 2019

Dunmall G (2016) The resurrection of Nantes: how free public art brought the city back to life. The Guardian. https://www.theguardian.com/cities/2016/jul/12/resurrection-nantes-france-how-free-public-art-brought-city-back-to-life. Accessed 13 Mar 2019

EY (2015) Cultural times: the first global map of culture and creative industries. EYGM Limited, Paris

Farrag NM, Abouhadid MA (2018) The viability of adaptive reuse of historic buildings as schools in Egypt. Curr Sci Int 7(3):337–343

Fernandez-Anez V, Fernández-Güell JM, Giffingerc R (2018a) Smart city implementation and discourses: an integrated conceptual model. The case of Vienna. Cities 78:4–16. https://doi.org/10.1016/j.cities.2017.12.004

Fernandez-Anez V, Valazquez G, Perez-Prada F, Monzón A (2018b) Smart city projects assessment matrix: connecting challenges and actions in Mediterranean region. J Urban Technol:1–26. https://doi.org/10.1080/10630732.2018.1498706

Ferrer J-R (2017) Barcelona's smart city vision: an opportunity. J Field Actions 16:70–75

Fonseca X, Lukosch S, Brazier F (2018) Social cohesion revisited: a new definition and how to characterize it. Innov Eur J Soc Sci Res:1–23. https://doi.org/10.1080/13511610.2018.1497480

Franco G (2018) Smart cities and historical heritage. TECHNE Special Issue 1:158–165

Garcia BM, Piccio L (2018) Why cultural heritage matters for urban resilience. The World Bank. http://blogs.worldbank.org/sustainablecities/why-cultural-heritage-matters-urban-resilience. Accessed 13 Mar 2019

Haans RFJ, Witteloostuijn AV (2016) Expected job creation across the cultural industries: a sectoral division and its implications for cultural policy. Int J Cult Policy 24(1):45–67. https://doi.org/10.1080/10286632.2015.1128420

Hassan C (2016) Is this the bottom? Digital finally surpass physical worldwide. https://www.digitalmusicnews.com/2016/04/12/ifpi-digital-revenue-overtakes-physical-for-the-first-time-ever/. Accessed 17 Mar 2019

Hegazy SM (2015) Conservation of historical buildings – the Omani-French museum as a case study. HBRC J 11(2):264–274. https://doi.org/10.1016/j.hbrcj.2014.03.010

Hellmuth E (2017) New study on arts as economic driver in St. Louis Region. Arts and Education Council. https://keeparthappening.org/blog/ae-news-and-events/new-study-arts-economic-driver-st-louis-region. Accessed 18 Mar 2019

Hernandez P (2018) Streaming now accounts for 75 percent of music industry reveneu. The Verge. https://www.theverge.com/2018/9/20/17883584/streaming-record-sales-music-industry-revenue. Accessed 17 Mar 2019

Hong Kong Economic TaCO (2018) Cultural matters. https://www.hketco.hk/en/culture/creative/index.html. Accessed 25 Mar 2019

Intellectual Property Watch (2016) Brief: UNESCO report on globalisation of cultural trade. http://www.ip-watch.org/2016/03/15/unesco-report-on-globalisation-of-cultural-trade/. Accessed 14 Mar 2019

Ivan (2022) Gaming statistics. Available at https://truelist.co/blog/gaming-statistics/. Accessed 20 July 2022

Jarus O (2013) Timbuktu: history of fabled center of learning. Live Science. https://www.livescience.com/26451-timbuktu.html. Accessed 18 Mar 2019

JCDecaux (2014) Enhancing urban heritage with technology. https://www.jcdecaux.com/fr/mobility-trends/enhancing-urban-heritage-technology. Accessed 17 Mar 2019

Jung TH, Dieck MCT (2017) Augmented reality, virtual reality and 3D printing for the co-creation of value for the visitor experience at cultural heritage places. J Place Manag Dev 10(2):140–151. https://doi.org/10.1108/JPMD-07-2016-0045

Kaczmarska E, Kaczmarska M (2018) Visual transformations within the space of the city. The Bilbao effect. Archit Urban Plan 7:35–49. https://doi.org/10.4467/2353737XCT.18.098.8793

Kaneda T, Greenbaum C, Kline K (2020) 2020 world population data sheet. Population reference bureau. [Online]. Available: https://bit.ly/3tJgT51

Klimovský D, Pinterič U, Šaparnienė D (2016) Human limitations to introduction of smart cities: comparative analysis from two CEE cities. Transylv Rev Adm Sci 47:80–96

Labadi S, Giliberto F, Rosetti I et al (2021) Heritage an the sustainable development goals: policy guidance for heritage and development actors. International Council on Monuments and Sites – ICOMOS, Paris

Lavarone R, Alberico L, Gravagnuolo A, Esposito De Vita G (2019) The role of cultural heritage in urban resilience enhancement. In: Calabrò F, Della SL, Bevilacqua C (eds) New metropolitan perspectives: smart innovation, systems and technologies. Springer, Cham

Lazaroiu GC, Roscia M (2012) Definition methodology for the smart cities model. Energy 47(1):326–332. https://doi.org/10.1016/j.energy.2012.09.028

Lekaota L (2018) Impacts of world heritage sites on local communities in the Indian Ocean Region. Afr J Hosp Tour Leis 7(3):1–10

Lewis R, Arthurs K, Berker M, Bishop A, Louis T, Slack J, Stennings S, Thomas H, Thomas I (2018) Cultural heritage for inclusive growth. British Council, London

Liberato P, Alen E, Liberato D (2018) Smart tourism destination triggers consumer experience: the case of Porto. Eur J Manag Bus Econ 27(1):6–25. https://doi.org/10.1108/EJMBE-11-2017-0051

López AM (2022) Guggenheim Museum Bilbao: annual attendance 2010-2021. Available at https://www.statista.com/statistics/781335/annual-number-of-visitors-in-the-museum-guggenheim-in-bilbao/. Accessed 20 July 2022

Mansfield C (2017) Travel writing in place branding – a case study on Nantes. J Tour Herit Serv Mark 13(2):1–7

Marsella S, Marzoli M (2017) Smart cities and cultural heritage. In: 2017 IEEE 14th International Conference on Networking, Sensing and Control (ICNSC), Calabria, Italy, 16–18 May 2017. IEEE

Mazzetto S (2018) Evaluation of the impact of sustainability principles on adaptive reuse interventions: the case of Qatari heritage experience. Archit Res 8(4):129–137. https://doi.org/10.5923/j.arch.20180804.03

Milmo C (2015) London's new skyscrapers 'inflict serious harm' on capital's historic landscape, heritage watchdog warns. Independent. https://www.independent.co.uk/arts-entertainment/architecture/londons-new-skyscrapers-inflict-serious-harm-on-capitals-historic-landscape-warns-heritage-watchdog-a6767296.html. Accessed 12 Mar 2019

Mortara M, Catalona C, Belloti F, Fiucci G, Houry-Panchetti M, Petridis P (2014) Learning cultural heritage by serious games. J Cult Herit 15(3):318–325. https://doi.org/10.1016/j.culher.2013.04.004

Murray W (2012) Selling tall: the branding and marketing of tall buildings. Council on Tall Buildings and urban Habitat (CTBUH), Shanghai

NatalinaCarrà (2016) Heritage/culture and social cohesion in the project of metropolitan city. Procedia Soc Behav Sci 223:583–589. https://doi.org/10.1016/j.sbspro.2016.05.353

Newman P (2020) COVID-19, cities and climate: historical precedents and expected transitions for the new economy. Urban Sci 4:32. https://doi.org/10.3390/urbansci4030032

Newman P, Kenworthy J (1999a) Costs of automobile dependence: global survey of cities. Transp Res Rec 1670(1):17–26

Newman P, Kenworthy J (1999b) Sustainability and cities: overcoming automobile dependence. Island Press, Washington, DC

Nicolas A (2014) Nantes' urban project: putting the Bilbao model to the test. Metro Politics.org. https://www.metropolitiques.eu/Nantes-urban-project-putting-the.html. Accessed 15 Mar 2019

Oad S, Mahesar RA, Lahbar GM, Jinliang Q, Shah SBH, Jingyi D (2018) Current status of cultural and creative industry (CCI). Appl Soc Sci Int Res J 1(1):12–22

OECD (2018a) Cultural & creative industries (CCIs): fulfilling the potential. Paper presented at the Unleashing the Transformative Power of Culture and Creativity for Local Development, Venice, Italy, 6–7 December 2018

OECD (2018b) Culture and local development: a framework for creating capacities at the local level. Venice

Otis Collage of Art an Design (2018) 2018 Otis report on the creative economy reveals California outpaces nations

Pietro LD, Mugion RG, Renzi MF (2018) Heritage and identity: technology, values and visitors experiences. J Herit Tour 13(2):97–103. https://doi.org/10.1080/1743873X.2017.1384478

Plaza B, Haarich SN (2013) The Guggenheim Museum Bilbao: between regional embeddedness and global networking. Eur Plan Stud 23(8):1456–1475. https://doi.org/10.1080/09654313.2013.817543

Plöger J (2008) Bilbao City report, CASE reports, vol 43. London School of Economics and Political Science, London

Pruetz R (2017) Lasting value: open space planning and preservation successes. Routledge, London

Psomadaki OI, Dimoulas CA, Kalliris GM, Paschalidis G (2018) Digital storytelling and audience engagement in cultural heritage management: a collaborative model based on the Digital City of Thessaloniki. J Cult Herit. https://doi.org/10.1016/j.culher.2018.07.016

Ramos F, Trilles S, Torres-Sospedra J, Perales FJ (2018) New trends in using reality apps for smart city contexts. Int J Geo-Inf 7:478. https://doi.org/10.3390/ijgi7120478

Riganti P (2017) Smart cities and heritage conservation: developing a smart heritage agenda for sustainable inclusive communities. Archnet-Ijar Int J Archit Res 11(3):16–27. https://doi.org/10.26687/archnet-ijar.v11i3.1398

Rosen G, Charney I (2016) Divided we rise: politics, architecture and vertical cityscapes at opposite ends of Jerusalem. Trans Inst Br Geogr 41(2):163–174. https://doi.org/10.1111/tran.12112

Sepasgozar SME, Hawken S, Sargolzaei S, Foozanfa M (2019) Implementing citizen centric technology in developing smart cities: a model for predicting the acceptance of urban technologies. Technol Forecast Soc Chang 142:105–116. https://doi.org/10.1016/j.techfore.2018.09.012

Sev A, Ozgen A, Basarir B (2011) Design criteria for high-rise buildings in historic cities: the case of Instanbul. Paper presented at the CTBUH 2011 World Conference, COEX, Seoul, Korea

Siew G, Allam Z (2017) Culture as a driver for sustainable urban development. Paper presented at the UIA 2017 Seoul World Architects Congress, Seoul

Soutinho P (2018) Cities' challenges in cultural heritage management. Urbact

Statista (2019) Annual evolution of the number of visitors to the Guggenheim Museum Bilbao (Spain) from 2010 to 2017. Statista. https://www.statista.com/statistics/781335/annual-number-of-visitors-in-the-museum-guggenheim-in-bilbao/. Accessed 13 Mar 2019

Tangbin ZD (2012) Analysis on relationship between skyscrapers and tourist cities' space. Atlantis Press, Paris

Teferi ZA, Newman P (2017) Slum regeneration and sustainability: applying the extended metabolism model and the SDGs. Sustainability 9(12):1–16

The World Bank (2018) Achieving the twin goals using cultural heritage, urban regeneration and sustainable tourism: lessons learned from Seoul. https://www.worldbank.org/en/news/feature/2018/02/07/achieving-the-twin-goals-using-cultural-heritage-urban-regeneration-and-sustainable-tourism-lessons-learned-from-seoul. Accessed 12 Mar 2019

Toronto (2017) Toronto designated a UNESCO creative city of media arts. Toronto. http://wx.toronto.ca/inter/it/newsrel.nsf/7017df2f20edbe2885256619004e428e/039943165 78b691c852581ca0071a914?OpenDocument. Accessed 25 Mar 2019
UCLG (2018) Culture in the sustainable development goals: a guide for local action, Culture 21. United Cities and Local Governments
UN Habitat (2018) Brief on world cities day 2018. UN Habitat, Shangai
UN Habitat (2020) Global state of metropolis report. Published in Nairobi. https://unhabitat.org/sites/default/files/2020/09/gsm-population-data-booklet-2020_3.pdf
UN Habitat (2022) Envisaging the future of cities. https://unhabitat.org/wcr/#:~:text=World%20Cities%20Report%202022%3A%20Envisaging,ways%20that%20cities%20can%20be
UN Habitat III (2016) New urban agenda. United Nations, Quito
UNCTAD (2018) Creative economy outlook: trends in international trade in creative industries. United Nations Conference on Trade and Development
UNESCO (1998) Action plan on cultural policies for development. Paper presented at the The Intergovernmental Conference on Cultural Policies for Development, Stockholm, Sweden, 2rd April 1998
UNESCO (2009) Investing in cultural diversity and intercultural dialogue, UNESCO world report. United Nations Educational, Cultural and Scientific Organization, Paris
UNESCO (2016) Culture: urban future: global report on culture for sustainable development. UNESCO Publishing, Paris
UNESCO (2017a) Early Chicago skyscrapers. UNESCO. https://whc.unesco.org/en/tentativelists/6235/. Accessed 18 Mar 2019
UNESCO (2017b) What is meant by "cultural heritage"? http://www.unesco.org/new/en/culture/themes/illicit-trafficking-of-cultural-property/unesco-database-of-national-cultural-heritage-laws/frequently-asked-questions/definition-of-the-cultural-heritage/. Accessed 18 Mar 2019
United Nations (2015) Resolution adopted by the General Assembly on 25th September 2015, vol 70. United Nations General Assembly
United Nations (2016) The world's cities in 2016. Data Booklet, New York
Vecco M, Imperiale F (2018) Cultural heritage in times of armed conflicts in the Middle East: much more than material damage? Introduction. J Cult Herit 33:262–263. https://doi.org/10.1016/j.culher.2018.08.011
Vegheş C (2018) Cultural heritage, sustainable development and inclusive growth: global lessons for the local communities under a marketing approach. Eur J Sustain Dev 7(4):349–360. https://doi.org/10.14207/ejsd.2018.v7n4p349
Weiss TG, Connelly N (2019) Protecting cultural heritage in war zones. Third World Q 40(1):1–17. https://doi.org/10.1080/01436597.2018.1535894
Wilson CR (2017) Culture and the sustainable development goals: shaping perceptions and inspiring change. Culture 360, vol Online. Asia-Europe Foundation
World City Culture Forum (2018) World cities culture report Bloomberg Philantropics. World City Culture Forum, London
Yang J, Černevičiūtė J (2017) Cultural and creative industries (CCI) and sustainable development: China's cultural industries clusters. Enterpreneurship Sustain Issues 5(2):231–242
Yigitcanlar T, Lee SH (2014) Korean ubiquitous-eco-city: a smart-sustainable urban form or a branding hoax? Technol Forecast Soc Chang 89:100–104. https://doi.org/10.1016/j.techfore.2013.08.034
Yigitcanlar T, Kamruzzaman M, Buys L, Loppolo G, Sabatini-Marques J, da Costa EM, Yun JJ (2018) Understanding "smart cities": intertwining development drivers with desired outcomes in a multidimensional framework. Cities. https://doi.org/10.1016/j.cities.2018.04.003
Zarzycki A (2016) Augmented cities: revealed histories. New Jersey Institute of Technology, Newark
Zvolska L, Lehner M, Voytenko Palgan Y, Mont O, Plepys A (2018) Urban sharing in smart cities: the cases of Berlin and London. Local Environ:1–18. https://doi.org/10.1080/13549839.2018.1463978

Chapter 6
Smart Cities and the Urban Economy: How Smart City Can Integrate Good Urban Economies

Abstract As the Smart City concept gains traction worldwide through the adoption of the technologies inherent to the concept, the potential increase in urban economic performance through efficiency and liveability levels of cities is welcomed. However, the cost factor associated with the adoption of those technologies is seen as unaffordable to some cities. Financing is increasingly leading to public-private partnerships that can provide the kind of mechanism that is best for delivering smart systems with sustainability outcomes. However, there are growing concerns of data privacy with PPPs. The need for regulatory frameworks that prevent the misuse of data should pave the way to enabling Smart City economic development based on soft and hard infrastructure integration and partnerships.

6.1 Introduction

The concept of Smart Cities is seen by some as the 'silver bullet' for solving numerous urban challenges. Cities today serve as the political headquarters, religious headquarters, centres for higher learning facilities and institutions and centres for science and technologies while offering improved and quality healthcare services, increased employment opportunities and quality hospitality services. However, most urban areas are also characterised by high urbanisation rates and a rapid population increase; these two phenomena translate into challenges ranging from traffic, pollution, unaffordable housing and unsustainable consumption of available resources, social inequalities and urban sprawl, among many others. In the course of trying to surmount these challenges, the Smart City concept has been seen as being among the most potent and practical approach (Allam 2021a), and it has been embraced in many places. Silva et al. (2018) suggest that Smart City systems can optimise the available infrastructure to solve numerous challenges, which, in its absence, would be solved by the adoption of such strategies like the expansion of hard infrastructures, among others, which have proven futile in many occasions. A report by McKinsey and Company (2018a) highlights that opting for expansion of infrastructure, without Smart City systems, may take years, and the same challenge

Z. Allam, P. Newman, *Revising Smart Cities with Regenerative Design*, Cities and Nature, https://doi.org/10.1007/978-3-031-28028-3_6

can be done faster and, sometimes, more cost-effective. This is the fundamental idea behind the economic benefit of Smart City.

The Smart City technologies such as the Internet of Things (IoT), artificial intelligence (AI), machine learning, big data and cloud computing have been outlined in other chapters to show how they can help solve urban challenges such as transport, energy, water, waste and construction, as well as social infrastructure involved in health, recreation, tourism, emergency services and education (Allam and Dhunny 2019). These technologies yield high volumes of real-time and comprehensive data, which, when analysed, facilitates speedy, efficient and on-point solutions to a plethora of these challenges. The adoption of Smart City systems is known to lead to an increase in both efficiency and performance of cities and, thus, can impact directly on the urban economy and echo positively on the GDPs.

Sepasgozar et al. (2018) argue that the ability to solve numerous challenges like traffic, security and environmental issues, as well as helping issues related to housing and business registrations, among many other solutions, enabled by Smart City technologies, has had positive impacts on attracting business opportunities in many cities. For instance, some cities have improved on their tourism status, others have managed to improve on job creation and employment opportunities, and others have managed to improve existing infrastructures such as in the case of Bilbao (Guggenheim Museum) and Nantes (old ship assembly warehouses) to economically revitalise their urban fabric, leading to higher liveability standards while shaping unique cultural urban identities.

However, the smart technologies that are associated with this concept are often seen as expensive, particularly noting that they are usually proprietary technologies offered by large ICT corporations. Chang et al. (2018) attribute the costs, partly with what he argues to be the 'complex, complicated and chaotic nature of cities', which make the installation of the smart components quite cumbersome. Chamoso et al. (2018) suggest that most of the providers of the relevant Smart City technologies are business oriented and, in competition, and would endeavour to maximise their profits as well as edge out their competitors. McKinsey and Company (2018b) note that such companies like IBM and Cisco which are among the leading companies in the sale of hardware, software, digital services, platforms and solutions have managed to make substantial profits since they have the capacity to provide Smart City technologies.

One way this is done is through their systems being unregulated; hence, they are provided as highly individualised solutions. Yigitcanlar et al. (2018) suggest that most of the Smart City components and platforms that each provider sells are not standardised, in terms of either protocols or systems; hence, city managers are pressured to contract more than one service provider, making the exercise rather expensive. This is true, especially when the capital investment is compared to the time it takes for return on investments to be realised. A sizeable number of smart projects in cities do not necessarily translate to direct financial returns but are geared toward promoting other services and ultimately boosting economic growth (Sarkar 2018; Hamilton and Zhu 2018). Unfortunately, most urban areas do have the financial capabilities to finance such projects, especially to fund customisable solutions made to address local challenges and/or projects with no direct financial returns to their successful completion. Therefore, this encourages cities to partner with selected private companies in the form of public-private partnerships (PPPs) to encourage investments in order to ensure its adoption at local level (Fishman and Flynn 2018; Milenković et al. 2017).

Those kinds of direct partnerships, however, are often contested as they open avenues for unsustainable and even unethical practices. First, as noted above, most of the companies are business oriented and would maximise their returns whenever an opportunity arises (Lam and Ma 2018). Second, such companies are often given very attractive tax rebates and incentives that allow them to dominate local markets such that small local companies are locked out of the competition (Cruz and Sarmento 2017). This makes it hard for new entrants, especially small, local ones, as it means that such large corporations would continue being part of projects in terms of operation and maintenance, services that local companies could be made to offer at high costs when they would do it better and cheaper (Allam et al. 2018; Allam and Newman 2018a). Moreover, for small local companies to engage at a later stage in projects, they would still need those large corporations since they control the proprietary technologies, in most cases, for the platforms and systems on which the concept is built. Large corporations monopolising the market can also misuse data for private purposes without caring for the privacy or security of citizens. Hollands (2015) further notes that such data could even be used as a campaigning tool as corporations try to gain an economic edge in the pursuit market control.

These issues of privacy and control were outlined in other chapters and continue to grow as a shadow over Smart City outcomes. However, the economy of cities can also benefit from Smart City systems; hence, there is need for new urban economy models that are ethical, inclusive and sustainable. It is critical that local companies and small start-ups that would help in easing the burden of unemployment, have local solutions to many of the sustainability issues outlined and can provide cost-effective bids for implementing such projects can have the opportunity to sprout and grow (Söderström et al. 2014). The chapter explores how this can be achieved.

6.2 Urban Regeneration and Economic Growth

Cities around the world have been striving to achieve sustainability and resilience in various forms, and a majority are targeting to improve their economic fronts by adopting different strategies like embracing Smart City technologies and capitalising on their cultural heritage and by adopting regeneration programs. The latter is hailed for the role it plays in fostering economic growth, especially through investments and face lifting activities targeting public infrastructures that on their end render a positive economic environment (Iovino 2018; Križnik 2018). Those programs are deemed important even when other strategies are to be implemented as they touch on infrastructural development, which are core to any urban economic activity. For instance, regeneration is seen as a nexus to safety and security in the cities, and this helps in attracting more investment in the neighbourhoods as those are seen as being desirable environments to live, work and play (Serrano et al. 2016; Tulumello 2017). A case in point is the transformation that the regeneration program brought in the city of Bilbao when the Guggenheim Museum was introduced, leading to the uplifting of the urban fabric (Baniotopoulou 2001). It is observed that before the program was initiated, the city of Bilbao leered in insecurity, dilapidated infrastructures and low economic activities, among many other forms of negatives (Wünsch 2017). Nevertheless, as it stands today, the city is now an economically vibrant metropolitan area that attracts visitors globally. This celebrated fabric now hosts numerous businesses where both locals and foreigners are able to earn a living (Plaza and Haarich 2013). One notable aspect learned from this case is that regeneration programs, if properly done, can open doors for foreign direct investments (FDIs) while, at the same time, allowing other strategies to unfold. This has been seen in other cities such as the city of Nantes (Mansfield 2017), Singapore (Bosch et al. 2017; EDB Singapore 2017) and Lhasa in Tibet (Zhang and Wei 2017), among others, that have regeneration to thank for their current status.

In order for regeneration programs to happen, one needs to take the risk for the initial investment in public infrastructures, but those are often thought to be the duty of government, either local or national. This is so because most public infrastructures are capital intensive and do not always have direct, short-term returns but are mostly undertaken to boost other sectors that, when compounded, results in tangible economic growth (de Paepa et al. 2017). The risk factor arises as most urban infrastructures are exposed to challenges like climate change-related disasters and threatened by human activities like terrorism and wanton destructions by ill-minded individuals. The risk also arises as there is no guarantee that the intended outcomes would be achieved; hence, most such programs are left in the hands of governments, which usually have the capacity to undertake substantial financial commitments that are synonymous with such projects (Ebrahim and Tavakoli 2016).

In cases where governments are not in a position to successfully finance such projects, they usually have the facilities of securing financial supports from different sources in the forms of foreign aids or loans from both local and international lenders. Indeed, such financing models have been in play in many economies where, especially in emerging economies, those have provided unnecessarily financial strains as countries face difficulties in honouring the repayment of loans. Brednkamp et al. (2019) highlight that among some of the consequences of struggling to pay or ultimately defaulting the payments include such issues as governments forfeiting their strategic assets to the lenders while also affecting the societal liveability levels on the long run. They explain that in most cases, projects financed under such unsustainable loans eventually stall or, if complete, are not of the quality that they deserve. Reinhart (2019) further posits that, in most cases where governments are caught up in the maze of unsustainable loans, the underlying reasons are either instigated by the action of the lenders who sometimes fashion the contracts to favour their side or from the failure of the governments to contain such issues like corruption, embezzlement of funds, rent seeking behaviour or failure to have a concrete, long-term development agenda. Mustapha and Prizzon (2018) also attribute this to the trend where most economies, especially in emerging and developing countries, opt to seek financing from 'nontraditional official and private lenders', who, as noted above, are more business oriented and naturally would want to maximise their returns at the expense of the borrower. They explain that this often happen as these lenders are not always careful and transparent by ensuring information sharing with their clients is actualised in the right way and at the right time.

With the above consequences already active in a number of economies, the theme of financing urban regeneration projects is thus often seen as a complex endeavour even though it is viewed as fundamental for building both economic resilience for public institutions while supporting conducive business environments and societal liveability levels. Andersen (2018) posits that this complexity arises since it calls into action the need to balance between seeking external financiers and applying financial mechanisms that allow the governments to self-fund projects. When external assistance is sought, it leads to debts, which sometimes prompt increasing taxes and reduce spending on entitlement programs like education and healthcare to service these debts. The first scenario is deemed as optimal but can only be achieved with strong debt management programs in place as advised by Mustapha and Prizzon (2018), while the second scenario is said to have the potential to spur challenges like social and economic inequalities and also lead to stalled projects, since the financial burden they raise may not be weighed down by only relying on public financing. If the two scenarios are not checked, they ultimately lead to stifled economic growth; hence, the ultimate objectives of regeneration are lost. Though these challenges are eminent, adopting prudent financing mechanisms is paramount as the overall return from regenerated infrastructures has been found to have positive impacts on all aspect of the urban fabric.

6.3 Smart Cities and the Economy

The economic benefits of the Smart City assembly of technologies are apparent. This provides an attractive narrative to many seeking more vibrant economies and resonates even to smaller economies and developing countries. For those economies that have already implemented many of these technologies, the outcome is evident in various aspects of their urban fabric. In particular, numerous economic activities have been supported by these technologies in such sectors like tourism and hospitality (Allam 2021b), transport and communication sectors, art, culture and creative industry (Llillevold and Haarstad 2019; Lavarone et al. 2019) and financial sectors, among many others. These sectors have in turn created employment opportunities for locals and also allowed them to capitalise on their cultural advantages to set up businesses in varying scales (Zvolska et al. 2018; Yigitcanlar et al. 2018). Furthermore, Smart Cities indirectly contribute to local economic growth by facilitating increased efficiency in most parts of the economy including the very expensive health sector (Pham et al. 2017).

Smart city technologies, especially in existing cities and built environment, are said to be less intrusive since they can fully rely on digital infrastructures (Shelton et al. 2014) unlike other large projects that demand the installation of hard infrastructures, posing a lot of challenges to planners and urban designers (Caragliu et al. 2011). As outlined in Chap. 3, the new decarbonised economy has now recognised that the best technological options of solar, battery and electric vehicles are small scale and can be delivered quickly and easily. However, they must necessarily involve smart systems that can integrate them into the urban fabric at different scales and different parts of the city. This fact, however, does not take away their potential to increase efficiency and sound performance of the resulting urban economies.

Many cities in the lower income part of the world have been reeling in a myriad of challenges, especially for lack of reliable data that planners and other stakeholders can rely on to solve urban challenges. To improve on the urban economy and other issues, Smart Cities, as showcased by Quiñones et al. (2018), emphasise the importance of data. Every infrastructure installed for the next economy needs to focus on capturing and generating data, which are then analysed and shared in real time to the respective agencies, utilities and companies responsible for managing them. Cheng et al. (2018) highlight that this data allows for real-time solution for a myriad of challenges that confront the city. This allows for quick action to address each issue accordingly. Sorda (2018) argues that improvement in technologies such as machine learning will mimic the concept of neural plasticity which may enable prediction of happenings like the change in weather patterns that is likely to impact on most cities for 30–50 years to come. Application of other emerging technologies such as the 5G and anticipated 6G, the digital twins and artificial reality will further play significant roles in cities (Moore 2020; Guilford 2019; Allam and Jones 2021). These innovations in future adaptation to climate change will assist with the efficiency of the urban fabric which directly relates to the performance of the local economy (Smith et al. 2022; Allam and Newman 2018b; Yang et al. 2022).

The problem that constantly arises is how such data are associated with the wider issues of data collected by companies as the basis of advertising and control of ideas, even controlling elections. And the use of surveillance data for various impositions of human rights continues to be an issue of concern. Data is the tool of science and engineering and is critical for solving issues, and hence commentators like Anand and Navío-Marco (2018) suggest that data is needed for the core functions associated with resource management, waste management and environmental sustainability concerns. The values associated with these must be integrated with the values associated with privacy, human rights and managing the broader concerns of data.

Despite the numerous benefits that Smart City infrastructure promises to cities, especially on their economic front, funding such projects is often a challenge, especially in emerging economies. Notably, this is because there is a lack of financing models that are exclusively associated with Smart Cities and how the technologies can be integrated into their other key objectives, including net zero and other SDGs. Flynn et al. (2018) support this argument by noting that the newness of this concept means that they lack basic and universally agreed frameworks under which smart

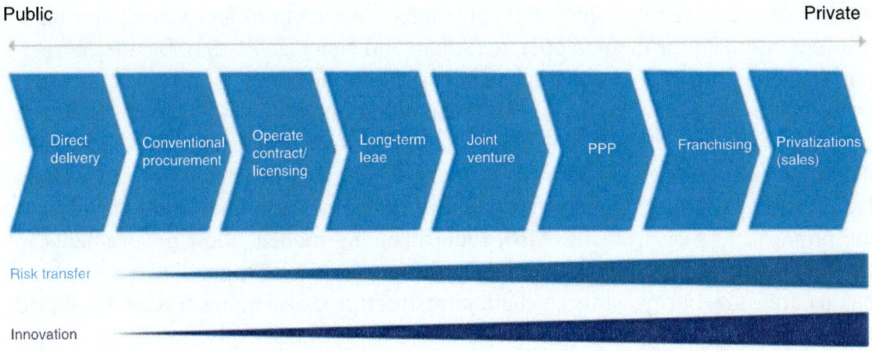

Fig. 6.1 Spectrum of procurement structures. (Flynn et al. 2018)

cities can be financed. They explain that, since most of the Smart City technologies demand modernisation of existing infrastructure or development of new infrastructure all together, a financial burden that is daunting for traditional financing models, most financiers would require a concrete conviction as to how they would get an attractive return on investment. In addition, with emerging economies, there are challenges of debt-to-GDP ratios, i.e. the amount of debt that the country owns both internal and external and where borrowings surpass its real GDP (hence, requiring a substantial amount of time to repay). Other problems with gaining finance include corruption, unclear procurement procedures, political instability and technical risks that require to be addressed so as to gain investor confidence for a financing model (Fishman and Flynn 2018).

Despite the shortcomings of not having a general financing model common for Smart City investments, emerging economies have the opportunity to capitalise on this very shortcoming to choose from the list of general financing models as advised in a Deloitte report (Flynn et al. 2018). These models include project finance, loans and leases, vendor finance, concessional finance, multilateral risk guarantees, multilateral credit guarantee and blended finance. On the same report, it is argued that local governments can opt to enter into procurements with private investors by adopting contracts that are best suited for Smart City infrastructures. The different types of contracts is as depicted in Fig. 6.1.

6.4 Financing Mechanisms: Public-Private Partnerships (PPPs)

The demand for Smart City and sustainability-based infrastructure in cities is so pressing and momentous that even economies that are not financially strong have been observed to go at lengths to ensure the upgrade of their urban infrastructure. To this end, cities have been observed to lure potential financiers by offering them attractive fiscal incentives that would guarantee sound returns and a conducive business environment for them to operate (Allam and Newman 2018a). Besides attracting financiers, the use of fiscal mechanisms in the uplifting of infrastructures has been observed to be a strategy being adopted by economies to encourage private investments in the public realm, hence allowing the governments to escape the burden of seeking financial support from lenders who may not necessarily offer loan facilities that would guarantee sustainability outcomes and high end returns from the projects (Liu et al. 2018). Also, such financing models allow governments to concentrate on other projects especially those with social aspects like education, healthcare and security, while key hard infrastructure is also being developed (World Bank Group 2018).

With those in place, locals and relatively smaller private companies and financiers may feel safer in taking the risk of committing resources into the projects, a case which is different when such partnership incentives are absent or meagre. McGreal et al. (2002) explain that fiscal mechanisms are also useful tools that have been used to ensure developments are not overemphasised on only one area or on sensitive areas like on Greenfields. Here, governments are said to deploy tax disincentives and penalties to areas developments are deemed as counterproductive and, at the same time, advance tax rebates, breaks and other incentives to such areas that require substantial attention. This would ensure even distribution of developments especially in terms of infrastructure. However, those have been often criticised as being inequitable and encouraging gentrification in the long run, when policies are only geared toward short-term planning (Morisson and Bevalicqua 2018).

Cities that utilise fiscal partnership mechanisms to facilitate implementation of projects, especially those that entail digital infrastructure, are said to require fewer financial commitments relative to stand-alone hard infrastructure. Soft (digital) infrastructures are the kind of Smart City infrastructure combined with net zero and other sustainability-based hard infrastructure outlined in this book. Thus, this partnership-based financing model could be the answer to financing and delivering Smart City and regenerative design implementation models.

Because digital infrastructure is based on the generation of substantial amounts of data, there are avenues regarding data management which are not clearly defined in contemporary legislation in many economies (Klievink et al. 2017; Malik 2013). Therefore, contracting private companies to undertake projects of such nature can be a catch-22 situation, especially if the company exploits the data systems for selfish profit-making agendas or for the sole purpose of gaining competitive advantages over its peers. However, the case of making sustainability outcomes much more the focus of such Smart City systems would suggest that these regulatory issues on data management must be given a high priority. This can include regulating to prevent the use of public data by foreign companies for international security purposes.

With these regulatory controls, it is possible to make fiscal mechanisms and PPPs work. One advantage of adopting either fiscal mechanisms or PPP, or a combination of both, is that governments are securing guarantees for the completion of projects, regardless of their sizes since most companies that have been observed to enter into such agreements have the financial ability, technical capacity and experience. These factors are paramount when looking at projects of national significance (Van Winden and van den Buuse 2017). It has also been observed that most companies that have the capacity to implement Smart City projects tend to retain an involvement in various capacities even when the projects are deemed as being completed. While this is portrayed as being the reason for lack of sufficient qualified personnel at local level to take over the project, it has been argued that those are encouraged so as to provide companies a longer time period to recoup on their investment (Collier and Cust 2015). This is a dangerous undertaking as it both hinders economic viability of the business community and offers avenues for monopolising data gathered from the numerous sensors across the Smart City networks. This would evoke the fear of privacy among citizens and would mean low acceptability of the concept as a whole. This calls for stringent regulatory measures to ensure such compromises are checked. Similarly, care should be extended to ensure that the cost of the projects is representative of the work, since the contracted partners may capitalise on fiscal tools to earn maximum revenues, thus making projects relatively higher than they would cost if financed under the linear financing model.

It is worthwhile to note that the two models are encouraged because the linear financing models cannot be sought since in the digital age, new legislation may be needed to support new infrastructure, and such moves would delay implementation of the projects. Similarly, legislation may be counterproductive to the implementation of innovative projects since those may rely on technologies that are in constant change, hence requiring extra financial commitments which may not be applicable and understood in linear financing models. Additionally, the diversity in technologies powering digital infrastructure such as Smart Cities is due to the availability of a myriad of platforms and systems that support different components. Legislation would need to incorporate each of these while allowing and encouraging investment to ensure an innovative urban environment.

6.5 Other Financing Mechanisms

Other financing mechanism used to enable infrastructure include tax increment financing (TIF) (Fishman and Flynn), municipal bonds and foreign direct investment (FDI).

The TIF scheme is majorly adopted where municipalities are unable to finance infrastructural development through the public financing option; hence, they divert expected, future tax revenues from a particular region into financing specific projects, thus compensating for the lack of direct public financing. Though this option has been embraced globally, it often falls short of expectation since the expected tax increment may not be certain and, also, the foregone tax could have been used in other sectors of the city. Also, Nguyen-Hoang (2018) notes that since most urban projects have a relatively long completion date, a given region where TIF is active would remain constrained of that tax, and this has the potential to affect other regions since their tax revenue increments are shared with the area under the project. Additionally, the particular regions may face urban decay themselves as the generated tax revenues are funnelled in particular regions or being redirected to other regions.

Municipal bonds on their part are loans that allow cities to raise financial tools pertinent to long-term projects that have the potential to constrain public budgets (Nallathiga 2017). The bonds are structured such that security provisions, maturity dates, repayment methods and redemption provisions are in favour of the municipality; hence, they can manage to complete the project without further strains. Though a potent option, critics argue that, in most cases and in the long run, bonds tend to be expensive as interest rates paid on the loans are dependent on current market value, where their uptake may be discouraged since they are capped at a given rate (Tigue 1995). They also demand that the municipality makes budgetary provisions for debt, which the municipality may not always be ready to commit.

Foreign direct investments or FDIs have also been used to finance projects, where municipalities receive financial support from foreign investors who invest directly on large-scale infrastructure projects. They rely on the government to be trustworthy enough, especially on debt repayment and management and the fundamentals of the economic potential of the project to be well developed. However, it can be a tool for geopolitical influence on policies and major infrastructural projects (Duanmu and Urdinez 2018).

6.6 Demonstrations, Special Economic Zones and Urban Development

In this section, we examine how government can assist in delivering integrated smart and sustainable technologies in soft and hard infrastructure-based projects such as net zero urban outcomes through two other mechanisms: demonstrations and special economic zones (SEZs).

Demonstrations are generally where a series of companies are encouraged to invest in innovative infrastructure and construction through the use of government land and/or the provision of grants, especially where partnerships can be created to do research on the projects, thus enabling greater certainty about the concept's mainstreaming potential.

One such demonstration in Perth, Australia, is WGV which has achieved net zero outcomes with all the SDGs ticked off and was set up to show how innovation in this area would require soft infrastructure from Smart City technology combined with hard infrastructure for energy through solar and batteries. Government land was available at the site, and grants were achieved from the federal, state and local governments partnering with Curtin University and the CRC in Low Carbon Living.

The WGV project is an infill residential development in a middle suburb called White Gum Valley. It has over 100 medium density housing units with different socioeconomic markets and was set up to demonstrate net zero carbon as well as other sustainability goals. Its intensive community engagement processes used the One Planet Living accreditation process to generate support locally and from all the agencies involved. It is therefore one of the first net zero precincts which is also delivering the UN Sustainable Development Goals. Solar photovoltaics and battery storage are incorporated into the development. A critical innovation was to test the viability of solar battery storage on strata buildings, and the project demonstrated the potential for a blockchain-based sharing system and potential for a new 'citizen utility' governance model, gaining attention from around the world (Green and Newman 2017; Eon et al. 2019).

The multiple sustainable development features such as water-sensitive design, energy efficiency, social housing, heritage retention, landscape and community involvement are aiming to provide inclusive, safe, resilient and sustainable living and have been assessed under the SDG framework.

The significance of innovation at WGV is that it demonstrated that a net zero carbon urban revitalisation project can do the following:

- Be commercially viable with all units selling very quickly.
- Build strong community acceptance of a higher density development through the provision of tree retention and a regenerated stormwater sump converted into public open space.
- Contribute to the Paris Agreement target that seeks to achieve deep decarbonisation while also delivering the United Nations SDGs.
- Build an integrated development using new green distributed technology, and support a first international demonstration of how to share solar energy through blockchain.

It is therefore a successful demonstration of how to do Smart City and sustainability through an integration of soft and hard innovations in infrastructure.

Special economic zones (SEZs) are another financing scheme that have been explored in different places of the world since 1959, with the Shannon zone in Ireland being the first to exploit this scheme (Zeng 2015). Initially, the scheme was said to have been highly utilised to encourage heavy industrial and ports activities, hence promoting growth in those areas (Adani 2017). The Shannon zone provides a classical case study to affirm this, since, at the time the zone was being constructed, the region was underdeveloped, but it is now reported to be Europe's busiest passenger and freight airport, with over 100 companies providing over 7000 job opportunities. It is therefore a successful economic project as it is reported that over €3 billion is generated each year from over 90% export activities (Shannon Chamber n.d.). After the success of this concept, it is said to have spread globally with countries like China, Singapore, Jordan and Mauritius, among others, with an aim of using an SEZ to attract FDI, job creation, foreign exchange and general economic growth, among other benefits (Kingombe and te Velde 2016).

In the last few years, the idea of SEZs has been gaining traction, where those encompass the development of technopoles as part of government to government Agreements. Klieibert (2017) showcases how SEZs have been integrated in spatial management in countries including the Philippines, where the concept is integrated in a gated community's project that is aimed at maximising the mixed use of available land in Metro Manila. Another notable case study is the Sino-Singapore Jilin Food Zone (SSJFZ) that was initiated to facilitate Singapore to attain its goal of having a high value and integrated agribusiness (R&D) and animal and pasture input production and processing facility (Tortajada and Hongzhou 2016; Rut and Davies 2018). Due to the land constraint in Singapore, the project is among the country's overseas farming projects that are enabled due to the collaboration between Singapore and China.

SEZs enable the application of special packages of fiscal incentives for development in a specific delineated zone (Allam and Newman 2018a; Allam 2018). The idea to fuse this type of approach to technopoles saw the increase in investment in high-tech urban development combining urban structure and digital infrastructure (Laboul and Croce 2014). An example is the case of Sino-Singapore discussed above and also the case of agro-processing and irrigation technologies adopted in some cities in countries like Israel and Egypt as explained by Gálvez-Nogales (2011). The author explains that governments in MENA (Middle East and North Africa) region which comprises 23 countries, for instance, are proactively offering incentives such as dedicated funds, business incubators and industry-specific infrastructures, among other things, to encourage the integration of agribusinesses in cities. Fiscal incentives are also becoming popular aimed at promoting ICT parks in different technopoles even in developing countries. Projects such as Konza City in Kenya (Konza Technopolis 2018), Eko Atlanta in Nigeria (Eko Atlantic n.d.) and Guangzhou Science City (UNESCO 2018) are among those that have benefited from fiscal tools applied to SEZ.

For countries and regions located within coastal regions, besides the SEZ, they could also exploit exclusive economic zones (EEZs) that have in the recent years been identified to possess significant value in terms of resources. Under the banner of Blue Economy, it is argued that cities within coastal strips have the capacity to exploit ocean resources, albeit in a sustainable way, to provide alternative sources of revenues and opportunities. Small Island Developing States (SIDS) in particular have been identified among economies that could greatly benefit from exploiting these resources as extra sources of economic and social benefits. A practical example is the Seychelles Sovereign Blue Bond, which raised approximately $15 million (Pouponneau n.d.). The capital raised allowed the country to commit sufficient resources for conservation as well as for climate action plans. With successful exploitation of EEZs, it would be possible to implement the Smart City agenda in coastal regions and in the SIDS as well as well as maintain the fidelity in sustainability agenda.

Building from this, numerous countries are now exploring ways and means of how to expand this approach of SEZ in the development of Smart Cities and sustainability. So far, there seems too little to demonstrate how smart and sustainability outcomes could be achieved through the SEZ mechanism. Its time this was changed.

6.7 Conclusion

Urban development is synonymous with economic growth, but the aspects of financing Smart Cities are known to be a challenging endeavour especially when the concept makes use of technologies that are in constant change and evolution, thus failing to fit into traditional legislation. This is particularly the case with privacy and surveillance data that are associated with Smart Cities. However, the chapter shows that economic development can be expanded with an integration of Smart City technologies and finance partnerships that are set up to enable strong sustainability outcomes. Such partnerships need to be regulated to prevent the misuse of data for control purposes that are not wanted by the cities and governments involved. There is thus a need to adapt both legal and financial legislation to encourage innovation in urban areas, as well as to capitalise on investments in short time periods. Other finance mechanisms can then be implemented and will be strongly assisted if there are demonstrations and special economic zones (SEZs) brought into the process of enabling urban development innovations. The integration of soft infrastructure (Smart City) and hard infrastructure (sustainable city) can be used to enable significant new urban development opportunities.

References

Adani (2017) Sustainability Report 2016–17. Reportno. Report Numberl, Date. Place Publishedl: Institution

Allam MZ (2018) Redefining the smart city: culture, metabolism and governance. Case study of Port Louis, Mauritius. Curtin University, Perth

Allam Z (2021a) Big Data, artificial intelligence and the rise of autonomous smart cities. In: The rise of autonomous smart cities. Palgrave Macmillan, Cham, pp 7–30

Allam Z (2021b) Global tourism and the risks of cultural homogeneity in smart and future cities. In: The rise of autonomous smart cities. Palgrave Macmillan, Cham, pp 75–90

Allam Z, Dhunny AZ (2019) On Big Data, artificial intelligence and smart cities. Cities 89:80–91

Allam Z, Jones DS (2021) Future (post-COVID) digital, smart and sustainable cities in the wake of 6G: digital twins, immersive realities and new urban economies. Land Use Policy 101:105201

Allam Z, Newman P (2018a) Economically incentivising smart urban regeneration. Case study of Port Louis, Mauritius. Smart Cities 1(1):53–74

Allam Z, Newman P (2018b) Redefining the smart city: culture, metabolism and governance. Smart Cities 1(1):4–25

Allam Z, Dhunny A, Siew G et al (2018) Towards smart urban regeneration: findings of an urban footprint survey in Port Louis, Mauritius. Smart Cities 1(1):121–133

Anand PB, Navío-Marco J (2018) Governance and economics of smart cities: opportunities and challenges. Telecomm Policy 42(10):795–799. https://doi.org/10.1016/j.telpol.2018.10.001. (In press)

Andersen CL (2018) Balancing the act: managing the public purse. Reportno. Report Numberl, Date. Place Publishedl: Institutionl

Baniotopoulou E (2001) Art for whose sake? Modern art museums and their role in transforming societies: the case of Guggenheim Bilbao. J Conserv Museum Stud 7:1–5

Bosch P, Jongeneel S, Rovers V, et al. (2017) City keys indicators for smart city projects and smart cities. Reportno. Report Numberl, Date. Place Publishedl: Institutionl

Brednkamp H, Hausmann R, Pienkowski A et al (2019) Challenges ahead. In: Abbas A, Pienkowski A, Rogoff K (eds) Soverign debt: a guide for economists and practitioners. Oxford University Press, London

Caragliu A, Del Bo C, Nijkamp P (2011) Smart cities in Europe. J Urban Technol 18(65–82)

Chamoso P, Gonzalez-Briones A, Rodríguez S et al (2018) Tendencies of technologies and platforms in smart cities: a state-of-the-art review. Wirel Commun Mob Comput 3086854:1–17

Chang DL, Sabatini-Marques J, da Costa EM et al (2018) Knowledge-based, smart and sustainable cities: a provocation for a conceptual framework. J Open Innov: Technol Mark Complex 4:5

Cheng S, Li X, Zhihan LV et al (2018) Data analytics of urban fabric metrics for smart cities. Future Gener Comput Syst 107:871–882. https://doi.org/10.1016/j.future.2018.02.017. (In Press)

Collier P, Cust J (2015) Investing in Africa's infrastructure: financing and policy options. Annu Rev Resour Econ 7(1):473–493

Cruz CO, Sarmento JM (2017) Reforming traditional PPP models to cope with the challenges of smart cities. Compet Regul Netw Ind 18(1–2):94–114

de Paepa G, Hart T, Long C (2017) Domestic resource mobilisation and the transition to sustainable development: synthesis of Asia case studies. Overseas Development Institute, London

Duanmu J-L, Urdinez F (2018) The dissuasive effect of U.S. political influence on Chinese FDI during the "Going Global" policy era. Bus Polit 20(1):38–69

Ebrahim Z, Tavakoli H (2016) Countercyclical financial instruments: building fiscal resilience to exogenous shocks. Commonwealth Secretariat, London

EDB Singapore (2017) Government leads digital disruption in Singapore. Available at: https://www.edb.gov.sg/en/news-and-resources/insights/innovation/government-leads-digital-disruption-in-singapore.html. Accessed 2 Nov 2018

Eko Atlantic (n.d.) An Investment opportunity on an unprecedented scale. Available at: https://www.ekoatlantic.com/about-us/. Accessed 18 Dec 2018

Eon C, Breadsell J, Morrison G et al (2019) Shifting home energy consumption through a holistic understanding of the home system of practice. In: Newton P, Prasad D, Sproul A et al (eds) *Decarbonising the built environment: charting the transition*. Palgrave Macmillan, Singapore

Fishman TD, Flynn M (2018) Part two: funding and financing smart cities series. In: Using public-private partnerships to advance smart cities. Deloitte Center for Government Insights, London

Flynn M, Rao AK, Horner J et al (2018) Smart cities funding and financing in developing economies: assisting development cites to finance their infrastructure gap through private sector participation approaches. Deloitte Development LLC, London

Gálvez-Nogales E (2011) The rise of agrifood technopoles in the Middle East and North Africa. Reportno. Report Numberl, Date. Place Publishedl: Institutionl

Green J, Newman P (2017) Citizen utilities: the emerging power paradigm. Energy Policy 105:283–293

Guilford G (2019) The coming 5G revolution. Available at: https://qz.com/1716218/to-build-out-5g-mobile-wireless-operators-must-spend-up-to-1-trillion-without-much-in-the-way-of-return/. Accessed 10 Mar 2022

Hamilton S Zhu X (2018) Funding and Financing smart cities. Reportno. Report Numberl, Date. Place Publishedl: Institutionl

Hollands RG (2015) Critical interventions into the corporate smart city. Camb J Reg Econ Soc 8(1):61–77

Iovino G (2018) Urban regeneration strategies in waterfront areas: an interpretative framework. J Res Didactics Geogr (J-READING) 1(7):61–75

Kingombe C, te Velde DW (2016) The role of special economic zones in manufacturing development in sub-saharan Africa: structural transformation and employment creation. In: Weiss J, Tribe M (eds) Handbook on industry and development. Routledge, London

Klieibert JM (2017) Exclusive development (s): special economic zones and Enclave Urbanism in The Philippines. Crit Sociol 44(3):471–485

Klievink B, Romijn B-J, Cunningham S et al (2017) Big Data in the public sector: uncertainties and readiness. Inf Syst Front 19(2):267–283

Konza Technopolis (2018) A smart city. Available at: https://www.konzacity.go.ke/the-vision/a--smart-city/. Accessed 18 Dec 2018

Križnik B (2018) Transformation of deprived urban areas and social sustainability: a comparative study of urban regeneration and urban redevelopment in Barcelona an Seoul. Urban Izziv 29(1):83–95

Laboul A Croce RD (2014) Private financing and government support to promote long-term investments in infrastructure. Reportno. Report Number|, Date. Place Published|: Institution|

Lam PTI, Ma R (2018) Potential pitfalls in the development of smart cities and mitigation measures: an exploratory study. Cities 91:146–156. In Press

Lavarone R, Alberico L, Gravagnuolo A et al (2019) The role of cultural heritage in urban resilience enhancement. In: Calabrò F, Della SL, Bevilacqua C (eds) New metropolitan perspectives: smart innovation, systems and technologies. Springer, Cham

Liu Y, Xu Y, Wang Z (2018) The effect of public target on the public-private partnership (PPP) residential development. Int J Strateg Prop Manag 22(5):415–423

Llillevold K, Haarstad H (2019) The deep city: cultural heritage as a resource for sustainable local transformation. Local Environ 24:1–13

Malik P (2013) Governing Big Data: principles and practices. IBM J Res Dev 57(3/4):1–13

Mansfield C (2017) Travel writing in place branding – a case study on Nantes. J Tour Heritage Serv Mark 13:2

McGreal S, Berry J, Lloyd G et al (2002) Tax-based mechanisms in urban regeneration: Dublin and Chicago models. Urban Stud 39(10):1819–1831

McKinsey & Company (2018a) Smart city solutions: what drives citizen adoption around the globe? Reportno. Report Number|, Date. Place Published|: Institution

McKinsey & Company (2018b) Smart Cities: digital solutions for a more liveable future. Reportno. Report Number|, Date. Place Published|: Institution

Milenković M, Rašić M, Vojković G (2017) Using public private partnership models in smart cities – proposal for Croatia. In: 40th international convention on information and communication technology, Electronics and Microelectronics (MIPRO). IEEE, Opatija

Moore D (2020) How 5G will power smart cities of the future. Available at: https://www.theglobeandmail.com/featured-reports/article-how-5g-will-power-smart-cities-of-the-future/. Accessed 11 Mar 2022

Morisson A, Bevalicqua C (2018) Balancing gentrification in the knowledge economy: the case of Chattanooga's innovation district. Urban Res Pract 1:1–21

Mustapha S Prizzon A (2018) Africa's rising debt: how to avoid a new crisis. Reportno. Report Number|, Date. Place Published|: Institution|

Nallathiga R (2017) Municipal bonds as a source of finance for urban infrastructure. Nagarlok XLVII(3):57–72

Nguyen-Hoang P (2018) Is tax increment financing a Fiscal Bane or Boon? J Plan Educ Res 41:1–2. https://doi.org/10.1177/0739456X18774121

Pham T, Tran T, Phung D et al (2017) Predicting healthcare trajectories from medical records: a deep learning approach. J Biomed Inform 69:218–229

Plaza B, Haarich SN (2013) The Guggenheim museum Bilbao: between regional embeddedness and global networking. Eur Plan Stud 23(8):1456–1475

Pouponneau (n.d.) A Seychelles' first debt-for-nature swap for ocean conservation. Available at: https://panorama.solutions/en/solution/seychelles-first-debt-nature-swap-ocean-conservation. Accessed 18 Oct 2021

Quiñones E, Bertogna M, Ferrer AJ et al (2018) Big Data analytics for smart cities: the H2020 class project. In: 18 proceedings of the 11th ACM international systems and storage conference. ACM, Haifa, p 130

Reinhart CM (2019) Financial crises: past and future. Bus Econ 54(1):3–15

Rut M, Davies AR (2018) Transitioning without confrontation? Shared food growing niches and sustainable food transitions in Singapore. Geoforum 96:278–288

Sarkar S (2018) Urban scaling and the geographic concentration of inequalities by city size. Environ Plan B: Urban Anal City Sci 46(9):1627–1644. https://doi.org/10.1177/2399808318766070

Sepasgozar SME, Hawken S, Sargolzaei S et al (2018) Implementing citizen centric technology in developing smart cities: a model for predicting the acceptance of urban technologies. Technol Forecasting Soc Change 142:105–116. In Press

Serrano E, Larrañaga I, Morteruel M et al (2016) Urban regeneration as population health intervention: a health impact assessment in the Bay of Pasaia (Spain). Int J Equity Health 15:145

Shannon Chamber (n.d.) Shannon free zone at a Glance. Available at: https://www.shannonchamber.ie/about/about-shannon/shannonfor-business/shannon-free-zone/. Accessed 5 Apr 2019

Shelton T, Zook M, Wiig A (2014) The "actually existing smart city". Camb J Reg Econ Soc 8(1):13–25

Silva BN, Khan M, Han K (2018) Towards sustainable smart cities: a review of trends architectures, components, and open challenges in smart cities. Sustain Cities Soc 38:697–713

Smith H, Medero GM, Crane De Narváez S et al (2022) Exploring the relevance of 'smart city' approaches to low-income communities in Medellín, Colombia. GeoJournal 88:17–38. https://doi.org/10.1007/s10708-022-10574-y

Söderström O, Paasche T, Klauser F (2014) Smart cities as corporate storytelling. City 18(3):307–320

Sorda KR (2018) Artificial neural networks – the future of smart cities. Available at: http://www.itwebafrica.com/home-pagex/opinion/24431 8-artificial-neural-networks%2D%2Dthe-future-of-smart-cities. Accessed 1 Jan 2019

Tigue P (1995) Structuring and sizing the bond issue: how to design an optimal financing approach: a practitioner's guide. Government Finance Officers Association, Chicago

Tortajada C, Hongzhou Z (2016) Food policy in Singapore. Ref Mod Food Sci 2017:1–7. https://doi.org/10.1016/B978-0-08-100596-5.21083-4

Tulumello S (2017) The multiscalar nature of urban security and public safety: crime prevention from local policy to policing in Lisbon (Portugal) and Memhis (US). Urban Aff Rev 56:1–36

UNESCO (2018) Science parks in Asia. Available at: http://www.unesco.org/new/en/natural-sciences/science-technology/university-industry-partnerships/science-parks-around-the-world/science-parks-in-asia/. Accessed 4 Apr 2019

Van Winden W, van den Buuse D (2017) Smart city pilot projects: exploring the dimensions and conditions of scaling up. J Urban Technol 24(4):51–72

World Bank Group (2018) Public-private-patereship legal resource centre. Available at: https://doi.org/10.3846/ijspm.2018.5224. Accessed 5 Apr 2019

Wünsch S (2017) The museum that changed a whole city: Guggenheim Museum Bilbao turns 20. Available at: https://www.dw.com/cda/en/the-museum-that-changed-a-whole-city-guggenheim-museum-bilbao-turns-20/a-41013716. Accessed 5 Apr 2019

Yang C, Liang P, Fu L et al (2022) Using 5G in smart cities: a systematic mapping study. Int Syst Appl 14:200065

Yigitcanlar T, Kamruzzaman M, Buys L et al (2018) Understanding "smart cities": intertwining development drivers with desired outcomes in a multidimensional framework. Cities 81:145–160

Zeng DZ (2015) Global experiences with special economic zones: with a focus on China and Africa. Reportno. Report Numberl, Date. Place Publishedl: Institutionl

Zhang Y, Wei T (2017) Typology of religious spaces in the urban historical area of Lhasa, Tibet. Front Archit Res 6(3):384–400

Zvolska L, Lehner M, Voytenko Palgan Y et al (2018) Urban sharing in smart cities: the cases of Berlin and London. Local Environ 24:1–18

Chapter 7
Conclusions: Redefining Smart Cities

Abstract While Smart City approaches promises transformative changes in urban management, its weaknesses have been exposed, particularly during the COVID-19 pandemic, as most of the added advantages were not recognized by residents. The proponents of this concept, mainly large ICT corporations, have not provided integrated solutions to the deeper structural challenges facing urban areas. Most cities solve their problems when they are inspired by local demand, with different stakeholders' inputs being incorporated into any projects. Therefore, "from-the-shelves" solutions have not worked. Our book argues that the Smart City concept needs to be redefined to include regeneration, culture, and heritage conservation, inclusive economic development, renewable energy, circular economies, education, gender equality, and community development. This will ensure that the technology is the servant of the city and not the master, and that the hard issues of city management are addressed. The chapter suggests that the sustainability and climate resilience agenda could learn from the Smart City agenda, and the two concepts integrated to provide robust solutions to urban challenges. To achieve this, partnerships between local communities, local governments, and other stakeholders are essential to enable the integrated approaches missing in most Smart City projects.

The prospect of the Smart City concept as the solution to the myriad of challenges that confront urban areas has largely failed to materialise. In this book, the examples of the Sidewalk Toronto project and Songdo have been cited. In particular, the weaknesses of this concept in fulfilling the anticipated goals were exposed during the height of COVID-19, where, despite a sizeable number of cities having deployed a varied number of smart technologies, the residents did not or could not have recognised many added advantages than their counterparts in 'non-smart' conventional cities. However, this is not the only weakness that has been identified. Most of the challenges are associated with the perception that proponents of this concept – mostly large ICT corporations – have been trying to establish the concept as the foundation of what makes an 'inclusive, safe, resilient and sustainable city' as required by SDG11, but in reality, they did not do much more than increase digital

Z. Allam, P. Newman, *Revising Smart Cities with Regenerative Design*, Cities and Nature, https://doi.org/10.1007/978-3-031-28028-3_7

data capacity. They did not or perhaps could not provide integrated solutions to these deeper and structural challenges.

As expressed in the book, due to the quest for profit maximisation, hinged on the substantial market capitalisation of the Smart City concept, most ICT corporations have been intensifying their efforts in introducing digital solutions (mostly universal ones), in the hope that different cities would automatically buy ‘smart’ infrastructure and deploy them for a range of deeper issues. However, urban areas across different regions have been identified to be unique in terms of financial capacity, socioeconomic needs, environmental challenges, population and demographic dynamics, among other issues. Therefore, ‘from-the-shelves’ solutions have not worked. More partnerships and integrated approaches are needed.

Most cities solve their deeper and more structural issues when they are inspired by local demand, with the inputs of different stakeholders being addressed and incorporated into any projects involving long-term infrastructure investment. The UN Habitat, one of the most influential institutions in the urban arena, has been instrumental in advocating for locally sourced solutions. In particular, it has been advocating for adherence to the post 2015 New Urban Agenda frameworks, goals and accords. The main objective of the New Urban Agenda has been the pursuit of sustainability, resilience, adaptability and liveability. But such agendas are not expressly addressed in the Smart City concept, and in some instance, it could be argued this planning model instead reduces the financial resources that could be deployed in achieving those global objectives.

Smart cities could be seen as being a distraction as it promised much but did not deliver enough because it was trapped by industry competitive agendas.

The weaknesses of the Smart City concept notwithstanding the positive aspects associated with the project cannot be overlooked. As expressed in this book, the most potent pathway is to intertwine the Smart City transformative pathway with global agenda, such that as the digital data capacity is increased, the deeper problems are actually being addressed as part of the smart agenda. This means that the technology is not being the master but the servant of the city.

If this can be achieved, then the global population would also experience an improvement in liveability status and see that the hard issues of city management are being addressed. This notion is captured in the book as ‘redefining the Smart City agenda’, such that approaches like regeneration, culture and heritage conservation and inclusive economic development can be central to Smart City urban planning. Such dimensions included in the Smart City concept would mean that renewable energy, circularity of economies, education, gender equality and community development would be fully addressed as part of adopting the Smart City projects being offered by ICT companies.

There are also things that the sustainability and climate resilience agenda could learn from the Smart City agenda. The robust technology adoption approach emphasised in the Smart City concept is one of the tools that could be tapped in the integration of other planning elements that this book has identified. For instance, in regard to promoting the sustainability agenda, the massive data generated by different installed smart devices could be exploited to help understand different urban aspects, sectors and activities addressing, and contribution to, sustainability transitions. These data can be used to help dramatise the sustainability outcomes in cities, in particular the net zero city agenda that are now clearly wanted but need detailed design projects based on the best data. In this book, we have provided a number of examples that show how net zero objectives could be advanced at small scale, where projects like rooftop solar energy production and battery and electric vehicles are all very commercial but need to be integrated using smart technology. This can be done differently in different urban fabrics and different income regimes. These projects will be promoting most of the SDGs.

Redefining of the Smart City concept could further help most urban areas conserve and regenerate cultural heritage in a more robust way than is the case today. This is critical as most urban areas have been observed to lose their identity and heritage as new modernist approaches, e.g. high-rise developments replacing informal housing have not necessarily provided better sustainability outcomes and almost always have removed their community structures. The book has outlined how community heritage and built heritage can be conserved and regenerated using new technology and regenerative design approaches. Smart systems and sustainability-oriented infrastructure can be integrated into community and cultural heritage. Smart and sustainable technologies can digitise cultural urban elements ensuring there is a legacy for future generations. The additionality is how such conservation can help to popularise heritage and contribute to job creation. However, this would call for concerted efforts between local communities, local governments and other stakeholders engaged as such partnerships will enable the integrated approaches that have been missing in most Smart City projects.

All levels of government need to get behind the redefining of Smart Cities using regenerative approaches. However, the new technologies of regenerative design and development are very much more applicable when related to the needs at a local level. Thus, local governments should endeavour to adopt the redefined Smart City concept, opening opportunities for emerging trends to be seamlessly incorporated, with a full set of local partnerships helping to define the integration goals and processes. This includes issues like food and energy insecurity, erratic inflation and the resulting challenges in supply chains.

The COVID shutdown in cities enabled many to re-localise their services. As localised technology works best with Smart Cities integrated through regenerative design, there are big opportunities to enable concepts like the '15-Minute City' to be implemented in any new projects. Even the new technologies like metaverse and digital twins need to be given a sense of direction to achieve the deeper values of urbanism. New smart systems can be applied effectively using mid-tier transit like new technology trackless trams along transit activated corridors that create 15-minute cities around station precincts. This ensures factors like proximity, accessibility, diversity and ubiquity in urban areas; these factors are fundamental to good cities. Redefining the Smart City concept would be very critical in ensuring such ideas are incorporated and enabled to fulfil the ancient agendas of urban planning as well as the need to use smart, new technology.